50 Biology Ideas
You Really
Need to Know

JV Chamary

greenfinch

Contents

Introduction

What is life? Biology is the study of life, so before we explore biology's most important concepts, we probably need a rough idea of what 'life' actually is. Look it up in a dictionary, however, and you'll end up going round in circles. Definitions will use phrases like 'living things' (in other words, life), organisms (again, life) and 'animals and plants' (that's right, life!)

Biology is a science of exceptions, which helps explain why life is so hard to label. Take viruses, for instance: many biologists don't think a virus is alive because it can't reproduce outside a host cell. But that ignores cases like *Mycobacterium leprae*, an intracellular parasite that can't live independently either. It's not surprising scientists haven't agreed on one definition.

While physics has lots of laws, biology's only law is evolution. But reproduction requires genes and organisms are made of cells, too. So the first three chapters of this book explore these fundamental topics, before we turn to the origin of life – technically, chemistry – and the tree of life. Later chapters are split into four sections of increasing organization: genes (chapters 6 to 16), cells (17 to 24), bodies (25 to 40) and populations (41 to 50). Along the way, we humans get our own chapter, as do viruses – which brings us back to that big question.

There are two ways to define life: what it *has*, and what it *does* – features like cells, and processes like reproduction. I think viruses are alive, so let's say life 'has' a container (either a cell or viral coat), an individual 'does' replication (reproduction) and populations adapt to an environment through evolution by natural selection. So what is life? Here's my idea: a self-contained entity with the ability to replicate and adapt. It works, but it's hardly catchy. If you come up with a better definition after reading this book, I'd love to hear it.

JV Chamary

01 Evolution

Every organism past and present is related through evolution, descended from common ancestors. The change to that family over time is driven by genetic mutations and environmental adaptations, a process that has continued uninterrupted since the first life on Earth, producing the biodiversity we see today.

Life is one big family, and you are one leaf on an unimaginably large family tree. Humans are not descended from monkeys, but we are both primates – cousins. Our very distant relatives include everything from bacteria to birds, and every organism is descended from the great-greatest of grandparents, a community of simple cells that were the ancestors of all life on Earth. But while we are related through common descent, we end up different because any given population – a family, a species, the animal kingdom – can change over time. This is half of the theory of evolution, or as Charles Darwin called it, 'descent with modification'.

Mutation

Until the 19th century, people believed each kind of organism (species) could not change – they were fixed or immutable. Then, in 1809, French naturalist Jean-Baptiste Lamarck presented his case for 'transformism' or 'transmutation' of species. His book, *Philosophie Zoologique*, proposed that species change due to pressures in the environment. Lamarck was right about why organisms adapt, but wrong about how they do it, thinking that adaptations could be acquired in an individual's lifetime and passed from one generation to the next – the giraffe's neck grew longer because its ancestors had stretched to reach tall trees, for example.

Lamarck's theory, the inheritance of acquired characteristics, fell out of favour as scientists realized that body cells could not transmit traits. In 1883, German biologist August Weismann called this the germ-plasm theory: only reproductive cells such as sperm and egg carried hereditary information. Austrian monk Gregor Mendel, whose breeding experiments with pea plants (revealed in 1865 and rediscovered in 1900), proved that characteristics are inherited as discrete particles – what we now call genes.

Today, the word 'mutation' is associated with genetic mutations and their impact on an individual's features, such as metabolism and appearance. Mutations are the ultimate source of biological variation, and provide the raw material for nature to weed out organisms that are not well-suited to their environment. This is the second half of Darwin's theory of evolution – natural selection.

Adaptation

Darwin published *On the Origin of Species* in 1859, the book that describes life's diversity and the mechanism that drives populations to adapt to their environment: evolution by natural selection. The theory is often simplified to 'survival of the fittest', which is slightly misleading. First of all, 'fittest' obviously involves more than physical performance – in biology, fitness means the ability to survive and reproduce. Second, the environmental pressures driving Mother Nature to choose between individuals – such as competition for resources or mates – does not pick the best; it only filters out the worst. It is better to think of natural selection as 'killing the least fit'.

> From so simple a beginning endless forms most beautiful and most wonderful have been, and are being, evolved.
> Charles Darwin

Natural selection is the main force driving evolution forward, but it is not the only factor influencing how populations change. The opposite of natural selection is 'purifying selection', a process that prevents unnecessary change to things that already work fine, or: if it ain't broke, don't fix it. A mutation can also have such a minor effect on an individual that it is effectively hidden from selection, so the mutation's fate in the population's gene pool depends on chance or random 'genetic drift'. In the 1930s, population geneticists incorporated such ideas into the theory of natural selection to create the modern evolutionary synthesis, or 'Neo-Darwinism'.

Think of evolution as a car on a slight slope. The vehicle will roll slowly downhill, driven by reproduction and genetic drift. Nature can apply the breaks to stop (purifying selection) or accelerate by pushing the gas pedal to adapt (natural selection), fuelled by mutations and variation.

The theory of evolution

Part of the problem in understanding 'the theory of evolution' lies in differences between popular and scientific terminology. Biologists agree that evolution happens – it is fact, it is true – but can disagree on details of its underlying mechanisms: the theory. People confuse 'theory' with 'hypothesis' (a hypothesis is a testable prediction; a theory is the framework for ideas). Like any scientific theory, the details are constantly being refined – just as the theory of gravity is not solely based on Newton's law of universal gravitation, but has added Einstein's general theory of relativity. 'Evolution' is another confusing word. It means 'unrolling' – any gradual change – but is often used as a synonym for progress or development, which explains why science-fiction movies sometimes claim individuals can 'evolve'.

Nature's adaptations are so amazing that it can be hard to imagine how they could possibly form by multiple evolutionary steps. This produces incorrect interpretations like those of Christian philosopher William Paley, who in 1902 compared life's complexity to the intricate workings of a watch. This Creationist thinking has been rebranded Intelligent Design, which is built on faulty reasoning: basically, if scientists have not found a 'missing link' or transitional fossil between two species, or the layman does not understand how a natural feature might have evolved, then the explanation must be supernatural. These logical fallacies are known as 'argument from ignorance' and 'the God-of-the gaps'.

Looking at nature, it can also seem that species are perfectly suited to their environment. This leads to attractive 'just so' stories to explain characteristics, like a giraffe's long neck. The organisms around us are a legacy of past adaptation, not the present-day environment. So to understand life's features, you have to understand why they evolved in the first place. To quote an essay by geneticist Theodosius Dobzhansky: 'Nothing in biology makes sense except in the light of evolution.'

Intelligent design

Intelligent Design (ID) is the idea that living things are so complex that they must have been created by an intelligent designer, such as God or aliens. The concept uses two main arguments. 'Specified complexity' claims that biological information, which encodes patterns and features, is so incredibly complex that the probability it could evolve by chance is impossibly low. Unlike a scientific theory, it fails to make testable predictions that would prove it true or false, and instead uses algorithms to detect design in abstract examples. 'Irreducible complexity' states that certain biological systems are too complex to have evolved from simpler parts. One example is the flagellum, the whiplike tail that some bacteria use for motility, which is compared to a mousetrap. In both cases, if you reduce the system to any combination of its components, it will not work. The evolutionary explanation is that parts of a system can appear in a step-wise process. Some bacteria use parts of the flagellum to stick to surfaces or release proteins, for example.

The condensed idea
Populations mutate and adapt over time

02 Genes

Genes carry biological information from one generation to the next, and shape an organism's every characteristic, from inner metabolism to its outward appearance. A full set of genes – the genome – encodes the instructions for building an individual and influences their ability to grow, survive and reproduce.

What is a gene? A dictionary will give an informal definition like 'a unit of heredity that determines a characteristic'. This is how many of us understand the concept, which is why we might say beautiful people have 'good genes', sporting ability is 'in your genes' or that researchers have discovered 'the gene for' some feature or disease.

Different genetic variants are also 'genes', so journalists might label a hypothetical gene for intelligence as 'the genius gene' or 'the stupidity gene', depending on the angle of a news story. Scientists do the same thing: for example, fruit fly development is controlled by genes such as hunchback and wingless – named after the effect of mutations, not what they do normally. Some of the confusion over the nature of genes can be blamed on the fact the concept has changed considerably in the past 150 years.

Units of heredity

Mankind has been breeding animals and plants for desirable traits for thousands of years, but the right explanation for how features are inherited was only revealed in 1865. The science of genetics began with the Austrian-Czech monk Gregor Mendel, who studied how characteristics like flower colour and seed shape are transmitted between generations. His breeding experiments with pea plants provided statistical observations that allowed him to devise laws of inheritance, principles that implied that the 'elements' that determine features are separate particles, discrete units of heredity we now call the genes.

The gene went from abstract entity to concrete object in 1910, when American geneticist Thomas Hunt Morgan found a fruit fly with a mutation that changed eye colour from red to white. His breeding experiments showed that patterns of inheritance were linked to being male or female, which is determined by sex chromosomes, so

There is no 'nature versus nurture' debate among biologists. But arguments are exciting so journalists often present nature and nurture as opposing views, then imply that genes completely determine a characteristic. On the other hand, some psychologists claim that certain behaviour is entirely determined by your upbringing. The truth is often somewhere in-between. Take human obesity, for example: genes control your pre-disposition to putting on weight through genetic variants that determine energy metabolism and whether your body responds to physical activity (nature), but staying fit and healthy also means not eating too many calories and getting regular exercise (nurture). And so an organism's features and behaviour are almost always the result of an interaction between its genes and the environment – nature via nurture.

the chromosome must be the physical structure that carries genes. Morgan and his students went on to show that genes are located at a specific place on a chromosome, and so the gene became a physical object at a distinct 'locus'.

Chromosomes consist of two types of molecule: proteins and DNA (deoxyribose nucleic acid). Which is the genetic material? In 1944, the Canadian-American trio of Oswald Avery, Colin MacLeod and Maclyn McCarty demonstrated that non-virulent bacteria could be transformed into a deadly strain in the presence of DNA, but not other parts of cells, proving that DNA is the molecule that carries genes. Scientists had previously assumed that proteins were the genetic material because their chemical building blocks – amino acids – are more varied than the four nucleotide bases in DNA, making them a better candidate for encoding biological information. This thinking changed after DNA's structure was revealed by James Watson and Francis Crick in 1953, as pairing between bases in the double helix revealed a way to copy information. The gene became a physical molecule.

Protein-coding sequences

Proteins do most of the hard work in the body, from forming a cell's inner skeleton to serving as signalling molecules between tissues. Most importantly, many proteins are enzymes, catalysing the chemical reactions of metabolism that drive life. A gene's affect on an organism's characteristics – the phenotype – is not always visible, but is ultimately the result of how its genotype affects biochemical activity within cells.

> It seems likely that most if not all the genetic information in any organism is carried by nucleic acid – usually by DNA.
> Francis Crick

In 1941, by exposing bread mould to X-rays, American geneticists George Beadle and Edward Tatum showed that mutations caused changes to enzymes at specific points in a metabolic pathway. This led to the 'one gene, one enzyme' view (later 'one gene, one protein'), where the gene was seen as instructions for making a functional molecule. Specifically, the gene became the blueprint for a protein.

After solving DNA's structure, scientists began to decipher how its instructions are used by cells, translating the genetic code of DNA into the language of proteins. The first discovery, by Francis Crick and colleagues in 1961, showed that genes use three-letter words, or triplets. The next five years showed that each triplet was a code for making a specific amino acid in a protein chain. But before a sequence

The double helix

Genes carry biological information, which is encoded as a sequence of nucleotide bases (letters) in DNA. The beauty of DNA's double helix structure is not its spiral, but the complementary pairing between bases on the two strands. This allows each strand to be a template or backup copy for the other, making it ideal for carrying genetic instructions.

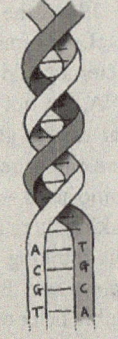

of DNA letters can be translated, it must be transcribed – read and copied – into messenger RNA (mRNA), so genes must encode an uninterrupted run of triplets: an 'open reading frame'. This reasoning led to the first gene being sequenced, from the bacteriophage MS2 virus, by Belgian biologist Walter Fiers in 1971.

American geneticist J Craig Venter led a team that published the first DNA sequence for a complete organism (the bacterium *Haemophilus influenzae*) in 1995 – the locations of potential genes was predicted by scanning the sequence for open reading frames. The genome was now data in a computer, and the gene became an annotated genomic entity.

Functional products

The protein-centric view of the gene is still the most popular way to explain its function, but DNA also encodes the blueprints for making RNA. Small 'transfer RNA' molecules are used to decipher the genetic code during translation, for example, whereas the machine that strings amino acids together into a protein – the ribosome – is built around 'ribosomal RNA'. Since the 1980s, various other kinds of 'non-coding RNA' have been discovered to control aspects of genetic activity.

While genomes in organisms like bacteria consist mainly of protein-coding genes, the genomes of many species are mostly non-coding DNA – around 98 per cent of the human genome does not encode proteins. The 'genomics age' has revealed that genes often consist of several pieces scattered along a chromosome, sometimes overlapping with each other. DNA contains functional elements, such as genetic control switches, that can be far from their associated gene. In 2007, Yale University biologists working on the ENCODE (Encyclopaedia of DNA Elements) project came up with a new, somewhat wordy definition: 'A gene is a union of genomic sequences encoding a coherent set of potentially overlapping functional products.'

The condensed idea
Units of heredity encode
functional biomolecules

03 The cell

The basic unit of life can function as an independent organism or form part of a multicellular body, and every cell is filled with various compartments that perform the countless reactions of metabolism. So it is somewhat ironic that cells were named after empty spaces.

In 1665, English polymath Robert Hooke published *Micrographia*, a collection of observations that he made using microscopes and telescopes. Amongst the many insects and astronomical objects is a detailed drawing and description of the honeycomb-like structure within a slice of cork. Hooke called those air-filled, empty spaces 'cells'.

Dutch microscopist Antonie van Leeuwenhoek was the first to see living cells, and from 1673 started reporting his finding in letters to the Royal Society in London. He described tiny moving particles and, assuming that motility meant animal life, concluded they were 'animalcules'. Van Leeuwenhoek discovered many microscopic organisms, including unicellular protists, blood cells, sperm and even the bacteria on tooth plaques, but progress then slowed until the 19th century, when optical microscopes and new tissue preparation techniques made it possible to peer inside cells.

Cell theory

The first to state that all life is made of cells was probably French plant physiologist Henri Dutrochet in 1824, but credit for the idea is usually given to two Germans who co-founded cell theory between them: botanist Matthias Schleiden and zoologist Theodor Schwann. In 1838, Schleiden claimed that every plant structure consists of cells or their products, while Schwann said the same applied to animals.

Schleiden and Schwann's cell theory had three tenets: all living things are composed of cells; the cell is the most basic unit of life; and cells form by crystallization. We now know the last one is wrong: cells do not arise by spontaneous generation from inorganic matter, but when a pre-existing cell divides in two, a process observed in algae by Belgian Barthélemy Dumortier in 1832, and in animal cells by Poland's Robert Remak in 1841.

Cell division was described in detail by German biologist Walther Flemming in 1882. Following the invention of oil-immersion lenses and dyes that clarified structures in the cell, Flemming used indigo to stain chromosomes and showed that they were copied and dragged into two daughter cells. This process, known as 'mitosis', is not carried out by all cells – only those whose chromosomes are contained in an envelope called the nucleus.

The nucleus

Scottish botanist Robert Brown is best known for describing the random movement of particles through a fluid – Brownian motion – but he also made big contributions to cell biology. In a paper read to the Linnean Society in 1831, Brown noted that 'a single circular areola ... or nucleus of the cell' could be found in various leaf tissues of orchids, suggesting that the structure was ubiquitous – and therefore important – in cells.

But the nucleus is not vital to life: bacteria are happy to leave their DNA – a circular chromosome and often a few 'plasmids' – floating naked in the cytoplasm. This can be an advantage as it allows a rapid

Germ theory

Today, we assume that diseases can be caused by germs invisible to the naked eye, but most people once believed maladies were caught by 'miasma' or contagion (pollution or direct contact). Dutch microscopist Antonie van Leeuwenhoek revealed organisms too small to see, but it was not obvious whether the microbes associated with an illness were a symptom or cause. Then in the 1850s, French chemist and microbiologist Louis Pasteur showed that beer, wine and milk contained cells that multiplied and caused food to spoil. Heating the liquids killed the germs, a treatment now known as pasteurization. Pasteur's tests helped disprove the idea that life arises from inorganic matter by 'spontaneous generation', leading him to reason that if microbes cause decay, they might also cause disease.

response to metabolic needs: genetic information is read from DNA and then interpreted to make useful proteins simultaneously, instead of reading (transcription) in the nucleus and interpreting (translation) in the cytoplasm.

Organisms are classified by whether their cells have a nucleus: eukaryotes have one, while prokaryotes do not – a distinction popularized by microbiologists Roger Stanier and C.B. Van Neil in 1962. Eukaryotes ('true nut' in Greek) include everything from the unicellular protists to multicellular organisms like animals and plants; prokaryotes ('before nut') include bacteria and archaea. So how did the nucleus arise? There are a dozen hypotheses, which come in two types: external origins involve a microbe evolving into the nucleus; internal origins suggest a cell folded its outer membrane inwards to form the nuclear envelope. External origin scenarios include a symbiotic relationship with one cell living inside another, an archean surrounded by a community of bacteria that later fused together, and infection by a complex virus.

Organelles

In 1884, German zoologist Karl Möbius described the reproductive structure in unicellular protists as 'organula' (little organ). The word 'organelle' is now used to describe any structure with a distinct function in eukaryotic cells. Many are even analogous to organs in the human body: mitochondria are like the lungs, breathing oxygen to release energy; the cytoskeleton resembles muscle and bone, providing movement and support; the plasma membrane is similar to the skin, a largely impenetrable barrier; and the nucleus is like the brain, except that it stores the memory of genetic ancestry rather than past experiences.

Prokaryotes have even littler organs. Whereas eukaryotic cells have subcellular compartments bound by one or more membranes, prokaryotic organelles are enclosed within protein-based shells. Some bacteria sense the Earth's

> If one compares the extreme simplicity of this astonishing structure with the extreme diversity of its innermost nature, it is clear that it constitutes the basic unit of the organized state; indeed, everything is ultimately derived from the cell.
>
> Henri Dutrochet

magnetic field using a chain of 'magnetosomes', for example, while others use 'carboxysomes' to concentrate the carbohydrate-making enzyme RuBisCO. Eukaryotic cells also have protein-bound compartments, mysterious mini-organelles of unknown function called 'vaults'.

Note that while eukaryotic organisms have complex cells and can form large, multicellular bodies, prokaryotes make up the majority of life on Earth. The oldest microfossils of eukaryotic cells are about 1.5 billion years old, but simple microbes were around for about two billion years before then. Complexity is not a measure of evolutionary success, and bigger is not necessarily better.

The condensed idea
The structural and functional unit of living things

04 The origin of life

Early in its history, our planet was a steamy, hellish world. Yet by 3.5 billion years ago, life had taken hold – as evidenced by cell-like fossil imprints in ancient Australian rock. So how did life on Earth develop from abiotic processes to developing key biological features such as genes, metabolism and a cell membrane?

In the 1920s, Russian biochemist Alexander Oparin and British mathematical biologist J.B.S. Haldane independently proposed that life originated in a primordial broth. Chemical reactions between simple molecules led to increasingly complex organic compounds in the ocean, perhaps powered by solar energy, to create what Haldane called a 'hot dilute soup'. The most famous test of this theory is the Miller-Urey experiment, when American chemist Stanley Miller, working in Harold Urey's lab at the University of Chicago, attempted to recreate conditions then thought to be have been present on the primitive Earth. In 1953, Miller added a mixture of inorganic gases – methane, ammonia, hydrogen and steam (no oxygen) – to a glass apparatus with a spark of electricity to simulate lightning. The final solution contained organic precursors like hydrogen cyanide, aldehydes and simple amino acids, but no polymers. Creating a dilute soup, it seems, does not quite recreate the necessary conditions for producing biomolecules – that requires a bowl to keep life's ingredients concentrated.

Cradles of creation

In 1871, Darwin wrote that he hoped the first organisms appeared 'in some warm little pond'. Numerous locations have since been suggested for where life began – some have argued that it started out in hot geothermal springs; others that it developed within pores on floating, beach-sized rafts of pumice stone created by volcanoes.

Many researchers, however, believe life originated underwater – partly because early Earth had no substantial continents, and also because rain would have diluted any soup in terrestrial pools. According to a leading theory, the cradle of creation was an alkaline hydrothermal vent similar to those found along the mid-Atlantic ridge, where superheated water rich in iron and sulfur bubbles up through the sea floor and minerals precipitate to form porous mounds of mineral. The surrounding water can reach boiling point, but is cool enough to support an ecosystem. According to British geochemist Michael Russell, who proposed the theory in 1997, vents would deliver two of life's requirements – energy and materials – to a single site.

Genetics or metabolism?

There is still no consensus about the first step on the path to life. Until the mid-20th century, many scientists thought proteins were genetic material. Oparin and Haldane both believed proteins encoded instructions for making organic droplets called 'coacervates' that replicated after assimilating other organic molecules through a primitive metabolism. Oparin thought genetic information was the first step to life, whereas Haldane believed a metabolic reaction arose first. Today, most researchers still fall into those two camps: genetics first, or metabolism first.

Disagreements boil down to how energy and materials are used. Genetics-first researchers say that all life replicates, so a prebiotic system must have encoded the instructions for making products – such as enzymes – to help bring materials together, enabling genes to copy themselves. Metabolism-first researchers argue that life is a process that consumes energy, so metabolic reactions are needed to harness power and assemble molecules.

Genetics-first supporters ask: how are materials assembled? Metabolism-first proponents ask: where is the energy to do that? The hydrothermal vent hypothesis is a metabolism-first scenario: seawater is more acidic than the alkaline fluid bubbling up from a vent, and this creates an electrochemical gradient between connected pores in a mound of minerals, so hydrogen ions (H^+) in acidic seawater flow down a concentration gradient towards the inside of the vent. Just as water pressure drives turbines in a hydroelectric dam, this gradient generates power that is captured by molecules between pores.

RNA world

One thing researchers agree on is that the first genetic system was not like the one we know today. Modern cells store instructions in DNA and use proteins to perform functions like catalysing reactions as enzymes. But because DNA makes protein, this creates a chicken-

According to the panspermia hypothesis, the seeds of life have been scattered across the Universe. In 1903, physical chemist Svante Arrhenius suggested that microbes might be pushed through space by solar radiation. Such naked transport is unlikely as genetic material would be destroyed, but delivery by interplanetary objects such as meteorites is theoretically possible, as dozens of terrestrial species have managed to survive trips into space, including bacteria and tiny animals called tardigrades. Most ideas are idle speculation: for example, 'directed panspermia' implies deliberate intervention by aliens, while astronomers Fred Hoyle and Chandra Wickramasinghe suggested some disease outbreaks came from space. The only hypothesis based on scientific evidence is 'pseudo-panspermia', where life is seeded by organic compounds and not whole organisms. Chemical analysis of objects such as the Murchison meteorite have uncovered fatty acids, amino acids, and nucleobases. One theory is that many of life's building blocks arrived during the Late Heavy Bombardment event around 4 billion years ago, when large asteroids regularly struck Earth.

and-egg paradox. This could be resolved by what lies within the ribosome, the molecular machine that cells use to synthesize proteins, where an enzyme-like 'ribozyme' made of RNA can be found. In 1982, American chemist Thomas Cech discovered ribozymes that work as independent catalytic RNA, and in 2002, molecular biologist Gerald Joyce made an RNA enzyme that can copy itself, enabling exponential growth and self-sustained evolution. This supported a proposal by British scientists Francis Crick and Leslie Orgel from the 1960s, that all prebiotic systems might have once been based on RNA – the so-called 'RNA World' hypothesis.

So why RNA and not another molecule? One clue came in 2009 from British chemists Matthew Powner and John Sutherland, who shone light onto a soup under 'prebiotically plausible' conditions. When exposed to ultraviolet radiation, the soup's ingredients were converted to cytosine and uracil, two of the four letters in RNA. This suggests the first genetic system originated from evolution by 'sunlight selection'.

Proto-cells

The cell is the basic unit of life, a compartment separating genes and metabolism from the environment. Modern cells are enclosed in a double-layered phospholipid membrane, but early 'proto-cells' probably used a bubble of fatty acids. Like oil droplets in water, fatty acids clump together and self-assemble into spherical bubbles: membranes are leaky to small molecules like RNA's building blocks but, once strung together, a long RNA molecule becomes too big to leave its bubble or 'proto-cell'. In 2004, Canadian biologist Jack Szostak found that the liquid inside a proto-cell becomes more concentrated and water follows by osmosis, causing the bubble to swell until it bursts and the fatty acids must reassemble. The

> Before the origin of life [organic substances] must have accumulated til the primitive oceans reached the consistency of hot dilute soup.
> J.B.S. Haldane

origin of cell growth and division might therefore be a result of physical forces, driven by self-replicating RNA.

Scientists can cook up prebiotic soups that recreate primitive conditions or uncover ecosystems that resemble basic processes, but we still might never know exactly how life began. Wherever its origin, at some point the first self-replicating bubbles left the comfort of their pores and became free-living cells – the first organisms.

The condensed idea
The transition from chemistry to biology

05 The tree of life

Evolutionary history is often depicted as a tree, with branches representing descent from common ancestors and its roots in the first cells. But the relationships between species, and especially among microbes, can be quite complex, suggesting that it might not be possible to represent all life in this way.

The first cells originated 3.5 to four billion years ago, but the founder of all current life on Earth – the last universal common ancestor, or 'LUCA' – probably resembled modern bacteria or archaea. One piece of evidence is that the genetic code is shared by all organisms. From these roots, however, the tree of life branches out to all species alive and dead, a powerful metaphor for evolutionary history – but is it right?

The ladder of life

Evolution is sometimes wrongly considered a scale of progress from primitive to perfect forms, with mankind as the pinnacle of creation. This idea stems from Aristotle and the *scala naturae* or 'great chain of being'. Around 350 BC the Greek philosopher arranged everything – both living and non-living – on a ladder, with rocks at the bottom and man at the top (the Bible would later knock us down a few rungs, putting us below God and the angels).

> In questioning the doctrine of common descent, one necessarily questions the universal phylogenetic tree. That compelling tree image resides deep in our representation of biology.
>
> Carl Woese

Aristotle was neither a creationist (believing life appeared suddenly), nor an evolutionist (assuming species arose through common descent). He was actually an 'eternalist' who thought that everything had always existed. Aristotle also believed that new life arose from non-living matter by 'spontaneous generation' – an idea based on observations he made, such as maggots emerging from meat. Spontaneous generation was finally disproved in 1859 by the germ theory of disease.

Ladders remained the dominant metaphor for over two millennia. In 1735, Swedish naturalist Carl Linnaeus published his *Systema Naturae*, in which he classified organisms using binomial names

Genetic material from one organism can sometimes be assimilated into the genome of another. In 1928, bacteriologist Frederick Griffith found that pneumococcus go from non-virulent to deadly after absorbing a 'transforming principle' (now known to be DNA) from dead bacteria. Such ,'horizontal gene transfer' is common among microbes and viruses, but rare in multicellular organisms. Most known cases of transfer occur between closely associated species, such as symbiotic partners or a parasite and its host, while genes donated by bacteria and fungi have ended up in various 'simple' animals, including sea sponges, insects and nematode worms. Transfers to vertebrate species seem especially rare – but DNA is regularly shuffled between genomes by viruses and other mobile genetic elements.

indicating a genus and species, such as *Homo sapiens*. This system established the field of taxonomy – grouping organisms by shared characteristics. Linnaeus also split nature into animal, vegetable and mineral. Trees of life appeared a century later, one famous example being 'Evolution of man' from 1874 – a great oak drawn by German biologist Ernst Haeckel. This 'tree', however, still implied a ladder of progression, with humans in the canopy.

The tree of life

On the Origin of Species contains a single illustration: a tree to represent 'descent with modification' (evolution). It has a 'V' shape with straight lines representing generations over time (descent) that split into branches (modification). Lines that extend to the canopy lead to living groups, the rest are extinct. Darwin's tree is not annotated with actual species, but naturalists were soon making trees of evolutionary history. Although Haeckel had represented human evolution as a ladder on his great oak, he was an early convert to Darwinian evolution and by 1866 depicted life in three parallel branches: plants, protists and animals.

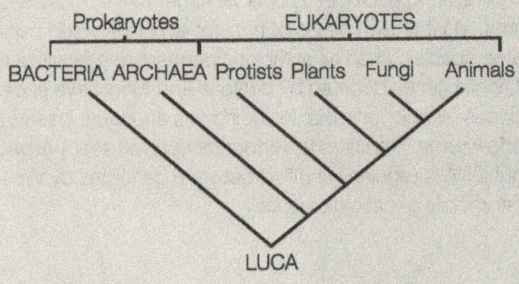

Phylogenetic tree of life

Scientists do not agree on a single tree of life. Here organisms have been grouped into six kingdoms (branches), three domains (upper case) and two empires (prokaryotes and eukaryotes). LUCA, the last universal common ancestor, was the most recent common ancestor of all living things.

Prokaryotes   EUKARYOTES

BACTERIA ARCHAEA Protists Plants Fungi Animals

LUCA

So how do you reconstruct life's family tree? Scientists today use an approach called cladistics (from *klados*, Greek for 'branch'): if groups share the same characteristics, you infer they have a common ancestor. The more features they share, the more closely related they are. Cladistics allows scientists to build a tree of phylogeny ('origin of races'). For extinct species, the only suitable characteristics are revealed by analysis of fossils. For living species, however, genetic information can be compared to spot the differences. In the mid-1970s, American microbiologist Carl Woese did this with RNA from ribosomes, the cell's protein-making machines. He noted that one group of microbes – the methanogens – lack an RNA fragment found in all bacteria, indicating they are a distinct group. Woese proposed that life be split into three 'urkingdoms' (now 'domains'): the eukaryotes (organisms with a nucleus) and two groups of prokaryotes – the bacteria and archaea.

Scientists do not agree on what a universal phylogenetic tree should look like. The three-domain system is widely accepted, but even the next taxonomic level – kingdoms – is debatable. Tree structures depend on which characteristics are compared, and researchers disagree on which are most relevant. Cladistics creates problems for taxonomy, too: in a phylogenetic tree, a properly 'monophyletic' branch or grouping should contain all the descendants of a common ancestor and no other organisms. Branches that ignore this rule are said to be 'paraphyletic'. Reptiles, for example, are a paraphyletic group because they exclude birds. But despite being warm-blooded, birds are descended from dinosaurs, which are reptiles. So birds *are* reptiles.

The web of life

Common descent involves passing down characteristics from one generation to the next by the 'vertical' transfer of genes, but an organism can sometimes acquire genetic material from sources besides its parents – so-called 'Horizontal gene transfer'. Based on this easy exchange of genes, American biochemist W Ford Doolittle claimed in 1999 that 'the history of life cannot properly be represented as a tree'. Horizontal gene transfer is relatively rare in multicellular eukaryotes, but the phenomenon seems to be common among prokaryotes: in 2008, for example, Israeli-German biologist Tal Dagan found that in 181 prokaryote species, over 80 per cent of genes had once been involved in horizontal transfer. So for prokaryotes at least, evolutionary history resembles a web.

So is there no such thing as *the* tree of life? It depends on what a branch represents. Darwin's drawings led to 'species trees' based on anatomy, but DNA lets modern biologists build 'gene trees'. If a branch then represents a genome inherited through vertical evolution, life's history has a rough tree shape: the trunk has the three domains (eukaryotes, bacteria and archaea) that run down to the base at the origin of life, while the twigs of horizontal gene transfer connect branches to form a web among prokaryotes. And if the first cells were anything like prokaryotes, then it is possible that our last universal common ancestor was not a single species, but a diverse community of gene-swapping microbes.

The condensed idea
Evolutionary history has branches but also webs

06 Sex

Birds do it. Bees do it. Even unicellular yeasts do it. But biologists do not get it. Despite being a more wasteful way to use resources in replication, the majority of complex organisms still prefer sexual reproduction to an asexual lifestyle. So why is sex so popular?

When you watch sexual intercourse on a wildlife documentary or at the zoo, the act is often unmistakeable, even in unfamiliar species. Sexual reproduction, however, is hard to define. Most simply, it combines genes from different individuals. This makes microbiologists happy because it means bacteria get to have sex, picking up DNA from each other, viruses and their surroundings. But many scientists prefer a narrower definition: sex is the union of two gametes, each carrying half a genome. If gamete cells are the same size then they are different 'mating types', as in yeast, but more generally, small gametes are sperm and larger cells are eggs.

Making gametes involves meiosis, the cell division that swaps DNA between paired chromosomes during recombination (see chapters 8 and 20), before separating those pairs into two cells. Two gametes fuse through fertilization, and an individual's sex (male or female) is based not on reproductive organs, but on whether they produce sperm or eggs.

Costs

The weirdest thing about sex is not *how* it happens, but why. When you think about it, asexual reproduction should be more common. Consider a human population where each couple produces two children, on average one boy and one girl. Now imagine a mutation causes a female to only make daughters, who reproduce the same way. These asexual mothers have twice the reproductive rate and would double each generation, eventually crowding out everyone else, leading to the extinction of sexual individuals, including all males. This disadvantage – sexual females produce half the number of offspring – is known as the 'twofold cost of sex'.

The twofold cost was proposed by American biologist George Williams in his 1975 book *Sex and Evolution*. Williams claimed it is a 'cost of meiosis', because each parent only contributes half its genes to

Squares represent males and circles are females. With sexual reproduction (left), assuming that each individual produces two offspring – one son and one daughter – then the population stays at a constant size. With asexual reproduction (right), numbers of asexual mothers rapidly increase to out-compete the sexual individuals. Males cannot reproduce on their own, so producing them is a waste of resources.

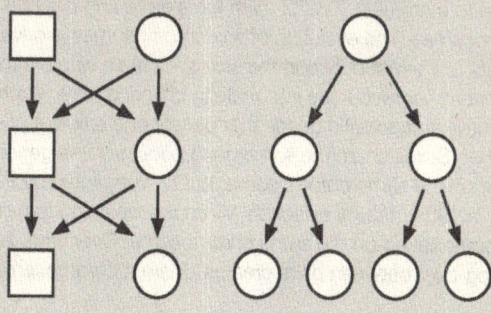

a gamete. Then in 1978, British scientist John Maynard Smith published *The Evolution of Sex*, saying that the cost is a 'cost of males', who cannot reproduce alone. More recently, Jussi Lehtonen, Michael Jennions and Hanna Kokko have argued that cost is influenced by the economics of parental investment: asexual females out-compete sexual individuals because they do not waste resources making males, who spend almost nothing on cheap sperm while females make large eggs. In a 2012 article, 'The Many Costs of Sex', the three evolutionary ecologists also point out that the cost is not twofold because investing in offspring is not just about producing gametes.

Benefits

To maintain sex in a population, its advantages must outweigh the costs of being out-competed by asexual females. In 1887, German zoologist August Weismann said sexual reproduction 'may be regarded as a source of individual variability, furnishing material for the operation of natural selection'. Geneticists later revealed that

Chromosomes and sex

An organism's sex is often determined by paired chromosomes. Most mammals have X and Y, females XX and males XY. The Y chromosome has a gene that controls whether male reproductive organs develop, so inheriting one X (XO) leads to femininity. Fruit flies use an XX/XY system too, but a different mechanism: sex is determined by the ratio of X to autosomes (non-sex chromosomes). Birds use a similar but opposite system to mammals, ZW/ZZ, with females having different sex chromosomes. The evolution of sex chromosomes is driven by conflicts of interest between the sexes – a male wants a mother to invest all her resources into making offspring, say, which leads to sexually antagonistic genes that benefit one sex at a cost to the other. Sexes share the same gene pool so these genes initially appear on the same chromosome, but as suggested by Ronald Fisher in 1931, natural selection will cause male and female genes to become linked on the same chromosome. Over time, less crossing-over between pairs creates different chromosomes.

variability is generated by mutation and recombination – the crossing-over of genetic material between pairs of maternal and paternal chromosomes to create new gene combinations.

In 1930, British population geneticist Ronald Fisher proposed that sex can bring together beneficial mutations from different parents via recombination. American geneticist Herman Müller had a similar idea in 1932, and in 1964 suggested that sex also stops bad mutations from building-up in DNA over time. Without recombination, 'Müller's ratchet' might even drive a species to extinction.

One theory for why sex is maintained is that it offers the benefit of interference, based on the Fisher–Müller hypothesis. Recombination separates linked genes – on the same chromosome, say – so that nature can detect their individual effects. If linked genes cannot be separated, a mutation in one gene – whichever has a bigger impact on an organism's fitness – would overshadow a mutation to the other.

Simply put: the first gene 'interferes' with natural selection's ability to act on the second gene. This effect is disrupted by crossing over between chromosomes, which separates genes and makes selection more effective so that species can adapt. But there are problems with this theory – for example, recombination not only brings good genes together, it also tears them apart.

Another dominant theory for sex is the benefit of parasite resistance. In 1978, John Jaenike argued that sex creates rare combinations of genes that could allow their carriers to resist parasites preying on hosts with more common genetic variants. Several field studies support the parasite resistance theory. For example, the New Zealand mud snail has mixed populations of sexual individuals and asexual females that produce daughters by parthenogenesis or 'virgin birth'. In 2009, Jukka Jokela, Mark Dybdahl and Curtis Lively found that asexual clones, which were common in the 1990s, had become more susceptible to fluke worms and decreased in numbers in under a decade. Although a handful of reptiles, amphibians and fish also use parthenogenesis, sex is the dominant form of reproduction in complex life. An estimated 99.9 per cent of animal and flowering plant species do it, suggesting that whatever the reasons for sexual reproduction, benefits outweigh the costs.

> Sexual reproduction does not have genetically favourable consequences in the short run sufficient to counterbalance the twofold advantage of not producing males.
> John Maynard Smith

The condensed idea
Sexual reproduction creates new gene combinations

07 Heredity

Characteristics are passed from one generation to the next according to common principles that apply to everything from plants to people. The discovery of these laws of inheritance revealed that information is transmitted as discrete particles, units of heredity we now call genes.

Darwin was right about many things but, like any good scientist, readily admitted to gaps in his knowledge. Most notably, he did not understand how characteristics were passed from parents to offspring. For example, why do some traits skip a generation, so you look more like a grandparent than either parent? As Darwin wrote in 1859, 'The laws governing inheritance are quite unknown'. Almost all naturalists at that time, including Darwin, believed an offspring's features – such as skin or fur colour – were a blend of those of its parents.

The blending theory of inheritance was proven wrong in 1865 by Austrian-Czech monk Gregor Mendel. During two meetings of the Brno Natural History Society, Mendel presented the results of an eight-year-long breeding experiment with pea plants. He and his assistants had used paintbrushes to carefully transfer pollen from the flowers of one plant to another plant (cross-fertilization) or to the same individual (self-fertilization).

Mendel studied seven characteristics: plant height, flower colour and position, pea shape and colour, and pod shape and colour. The first few years were spent self-fertilizing plants until they displayed the same features every generation. These pure breeds were crossed to create hybrids, which were then bred with one another, with Mendel noting the number of offspring with each trait at every step – observations that form the basis of principles that explain how biological information is inherited.

The laws of inheritance

When Mendel crossed pure-bred plants with different traits, offspring would resemble one parent, not a blend of the two. Crossing plants with round peas and those with wrinkly peas never produced semi-wrinkly peas. In fact, the offspring would inherit one trait but not the other. Crossing round with wrinkly always produced round-shaped

After crossing pure-bred plants with round-shaped peas and plants with wrinkly peas, all offspring had round peas, so the round trait was dominant. After crossing first generation (F1) hybrids with each other, some second-generation (F2) plants regained the recessive, wrinkly trait. This revealed the genes controlling characteristics are inherited as discrete units.

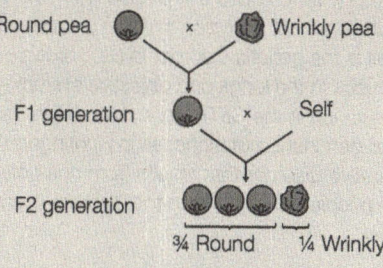

peas. Mendel realized that one of the two traits dominates, a relationship he represented using uppercase and lowercase letters. For pea shape, the round 'R' trait is 'dominant' and wrinkly 'r' is 'recessive'.

The first generation (F1) hybrids created by crossing pure-bred plants only displayed dominant traits such as round peas. But after Mendel crossed those F1 hybrids with one another, some of the second generation (F2) hybrids would recover the recessive trait from their original pure-bred grandparents. Mendel concluded that each characteristic is determined by 'elements' that come in pairs, one from each parent, and are separated before being transmitted to the next generation. This is the first principle of inheritance – the law of segregation.

After counting the number of F2 hybrids displaying each trait, Mendel found the ratio of plants with dominant traits versus recessive traits was always 3:1. This can be explained by elements behind a trait being inherited in pairs: the four possible pairs in the F2 generation are 'RR', 'Rr', 'Rr' and 'rr', three-quarters of which carry the dominant 'R'. Although 'Rr' plants have round peas, they also carry the 'r' element, so if crossed with each other, the 'rr' pair

Traits that follow Mendel's laws of inheritance, determined by a single gene with dominant and recessive alleles, are relatively rare. In humans, characteristics such as eye colour and tongue-rolling ability were once thought to be Mendelian, but are now known to be caused by multiple genes. One of the few Mendelian features is earwax: the allele controlling the wet earwax trait is dominant to dry, which occurs in people with two recessive alleles – a mutation most common in Asian populations. Another Mendelian trait is the genetic disorder cystic fibrosis – excess build-up of mucus in the lungs and digestive system – caused by a recessive mutation in the CFTR gene. Mendelian disorders can also be due to dominant mutations, as in Huntingdon's disease, a neurodegenerative disorder that results from one defective copy of the gene – people will suffer from the disorder even if the other allele is normal.

can reappear in their offspring. This explains why traits can skip a generation, so you have features that more closely resemble a grandparent rather than either parent.

Mendel also studied combinations of characteristics, such as crossing plants with green, round peas (both dominant traits) and plants with yellow, wrinkly peas (both recessive). These crosses not only produced offspring resembling their parents, but also created hybrids with new trait combinations, such as green, wrinkly peas and yellow, round peas. This suggested that when the elements behind a trait become separated, the elements for each characteristic remain separate, too. This is the second principle of inheritance – the law of independent assortment.

Genes

The 'elements' that determine a characteristic are now called genes and the variants behind different traits are alleles. The combination of allele pairs gives an individual's genotype – two copies of the same allele ('RR'/'rr') is said to be a homozygous genotype, different variants ('Rr') is heterozygous. Results of genetic activity are displayed in the phenotype. Rather than continuous variation predicted by blending, Mendel's experiments proved that units of heredity are discrete particles that do not blend. Today, we know these units as 'genes' – a term coined in 1909 by Danish botanist Wilhelm Johannsen.

It requires indeed some courage to undertake a labour of such far-reaching extent ... the importance of which cannot be overestimated in connection with the history of the evolution of organic forms.
Gregor Mendel

Breaking the laws

Mendel's work went largely unnoticed until it was rediscovered in 1900. But as biologists then replicated the monk's experiments, they found that while his seven characteristics produced the same results, others displayed inheritance that deviated from Mendel's laws. For instance in 1905, British geneticists William Bateson, Edith Rebecca Saunders and Reginald Punnett crossed pure-bred pea plants with purple flowers and long pollen grains against plants with red flowers and round grains. Purple and long traits are dominant, so only one in 16 second-generation (F2) hybrids should have had red flowers with round grains, but there were three times more than expected.

The British trio suggested that flower colour and pollen shape were somehow coupled, which would explain why certain combinations of traits tend to stay together over successive generations. The study of recombination would later reveal that traits are coupled because genes are physically linked on the same chromosome. Geneticists have found numerous exceptions to Mendel's laws, including co-dominant alleles and complex phenotypes determined by multiple genes. Either deliberately or through luck, Mendel's seven features obeyed his laws of inheritance, but others did not. Nonetheless, while simple 'Mendelian' traits are rare, the laws of inheritance still explain the underlying mechanism behind the transmission of biological information.

The condensed idea
Characteristics are inherited as combinations of genetic variants

08 Recombination

Genes are not passed between generations as discrete particles, but are arranged in a linear order along chromosomes. This not only explains why characteristics can be inherited together, it also allows offspring to inherit new combinations of genes from their parents by the shuffling process of genetic recombination.

The start of biology as a modern science can be traced back to one man, one lab, one date: American geneticist Thomas Hunt Morgan and the 'Fly Room' at Columbia University in April 1910. He wanted to study how hereditary traits influenced animal development and needed a species that would breed quickly and regularly. His search led to the fruit fly *Drosophila melanogaster*, which produces a new generation every 12 days. In a cramped lab smelling strongly of bananas, Morgan and his students would uncover principles of inheritance that Gregor Mendel missed, while proving that genes are located on physical structures – chromosomes.

Chromosomes

While prokaryotes such as bacteria carry a single, circular chromosome and extra structures called plasmids, the eukaryotes – organisms whose cells contain a nucleus – use linear strands. American zoologist Theophilus Painter counted human chromosomes under a microscope in 1923 and claimed we have 48 in total. Then in 1956, Joe Hin Tjio and Albert Levan corrected that figure to 46: one pair of sex chromosomes (usually XX/XY) plus 22 pairs of 'autosomes'. Humans and fruit flies are 'diploid', with matching pairs from each parent, but some species are 'polyploid', with multiple sets, and others are 'haploid', with one set.

Chromosomes, structures made of DNA and protein, were observed in 1878 by Walther Flemming, who tracked their movements during cell division. The possibility that chromosomes carried genes was proposed independently by German biologist Theodor Boveri and American geneticist and surgeon Walter Sutton. Boveri observed that sea urchin embryos needed chromosomes to develop normally. Sutton tracked how the structures behaved in grasshopper cells and suggested that sperm and egg carry pairs of chromosomes that

How do you pinpoint the location of genes on a chromosome? One of Morgan's students, Alfred Sturtevant, devised a method based on the way recombination affects the statistics of inheritance: if two genes are next to each other, there is almost zero chance of crossovers between them; if two genes are far apart then there is a 50-50 chance they will be separated. This is reflected in the combination of new characteristics: if a combination is always inherited together, its 'recombination frequency' is zero; if half of offspring have a new combination that differs from parents then the frequency is 50 per cent. By using recombination frequencies for any pair of traits, Sturtevant was able to calculate the relative distance between any pair of genes. He named the units of distance – the CentiMorgan – after his mentor, and used them to annotate fruit fly chromosomes with the location of any genetic trait with an observable, phenotypic effect.

undergo 'reducing division' (meiosis – see chapter 20), which is what allows an embryo to end up with pairs. Sutton's 1902 study concludes: 'the association of paternal and maternal chromosomes in pairs and their subsequent separation during the reducing division ... may constitute the physical basis of the Mendelian law of heredity'.

Linked genes

During his fruit fly studies Morgan hoped to find insects that had undergone a sudden change – a mutation. But his first two years proved, well, fruitless. Then, in 1910, he found a male with white eyes rather than the usual red. When this mutant was bred with red-eyed females, the offspring all had red eyes. This suggested the red allele was dominant and white recessive, obeying Mendel's laws of inheritance. When Morgan continued by crossing the red-eyed offspring, the second generation also had the expected Mendelian ratio of 3:1 for dominant to recessive traits, but with an important difference: some males had white eyes, but not the females.

Morgan realized that eye colour was linked to sex, which is determined by chromosomes. Fruit flies are diploid, with pairs of paternal and maternal chromosomes: males are XX, females are XY, with three pairs of autosomes each. Based on the patterns of heredity, Morgan showed that females would only be white-eyed if they inherited the trait from both parents, so the gene for eye colour must be linked to the X chromosome. More generally, this suggested that genes are located on a physical structure – the chromosome theory of inheritance.

By studying features such as eye colour, Morgan and his students discovered that many show a skewed pattern of inheritance, best explained by genes being on chromosomes: genes can be physically linked. If two genes are next to each other, there is a good chance they will be inherited together. If genes are at opposite ends of a chromosome (or on different chromosomes), they are more likely to become separated and follow Mendel's law of independent assortment. And when genes are separated, they can create new combinations of traits that did not exist in parents – recombination.

Crossover

Recombination occurs because of a physical crossing-over between pairs of similar, 'homologous' chromosomes. During meiosis, before a sperm or egg cell splits its pairs of genetic material in half, the homologous chromosomes line up side-by-side. At some points along

Crossover

Genes are located on chromosomes. Thomas Hunt Morgan compared this physical structure to beads on a string. Recombination is the result of crossovers between a pair of chromosomes while creating reproductive cells, separating combinations of parental traits.

the length of their structures, they come into contact, swapping genetic material before separating. In 1931, American geneticists Harriet Creighton and Barbara McClintock observed crossover under a microscope, during meiosis in maize (corn) plants, and found it corresponded to recombination of genetic traits.

That the fundamental aspects of heredity should have turned out to be so extraordinarily simple supports us in the hope that nature may, after all, be entirely approachable.
Thomas Hunt Morgan

Recombination creates genetic diversity between generations. Your parents inherited genes from their parents, but that information was not left untouched before it reached you: when your parents produced sperm or egg cells, matching sections of homologous DNA from your grandparents were exchanged to create a random, unique combination of traits.

Morgan visualized the process by comparing genes on chromosomes to beads on a string, but crossover is actually a biochemical reaction that involves cutting and pasting DNA. Chromosomes consist of two DNA strands that, when cut, either grow or replace each other at a cross-shaped 'Holliday junction'. The strands are complementary, so homologous recombination also allows cells to use one strand as a template when repairing breaks in chromsomes.

The breakthroughs that began in the Fly Room at Columbia University were recognized by the scientific community in 1933 when Morgan won the Nobel Prize in physiology or medicine. As well as showing that an abstract concept – genes – had a physical form, the Fly Room's use of statistics helped turn biology from a field that largely relied on describing features, to an experimental science that could rival chemistry and physics.

The condensed idea
Chromosomes can swap genetic information

09 Mutation

Variety is not just the spice of life, it is nature's main ingredient. The source of that variety is genetic mutation and, whether harmful or beneficial, DNA changes generate the majority of the variation that passes through the evolutionary sieve of natural selection.

The process of evolution became widely accepted by naturalists after Darwin published *On the Origin of Species* in 1859. But while many believed the book made a convincing case for 'descent with modification', they also doubted Darwin's mechanism for the driving force – natural selection. As a result, several competing theories were presented between the 1880s and 1930s, a period known as 'the eclipse of Darwinism'.

One alternative was the mutation theory, advocated by Dutch botanist Hugo de Vries, who helped rediscover Mendel's experiments with pea plants. Over 13 years breeding the evening primrose, de Vries discovered numerous variants that differed from pure-bred parents to the extent that they seemed like precursors to new species. Rather than a gradual process of change via natural selection, de Vries suggested that species originated in sudden jumps, or 'mutations'. Later studies found rearranged chromosomes in his mutant plants, but in the early 1900s the cause of mutations was not known.

Genetic variation

The word 'mutation' became associated with genetic changes after American biologist Thomas Hunt Morgan proved genes were located on chromosomes, in 1910. Early progress in genetics research was slow because natural mutations were rarely found. This sticking point was overcome by Morgan's former student, Hermann Müller, the first geneticist to induce mutations artificially. In 1927, Müller created over a hundred mutants by exposing fly sperm to heavy doses of X-rays, which recreated natural mutants like Morgan's white-eyed fruit flies, as well as other flies with novel phenotypes. Müller noticed that mutants had rearrangements in the linear order of genes on chromosomes and, while radiation was often lethal or caused sterility, many survivors transmitted mutations to subsequent generations.

Evolution was clearly fuelled by genetic variation, and by the 1930s one big question remained: did mutation occur before or after natural selection? In principle, selection could either act upon pre-existing mutations, or prompt an organism to mutate. In 1943, biologists Salvador Luria and Max Delbrück tested these alternatives by comparing the distribution of induced mutations and spontaneous mutations in colonies of bacteria.

DNA changes

Genetic mutations differ from DNA damage in one important way: relative stability allows them to be inherited. These changes to DNA can be large or small. The rearrangements seen in de Vries's primrose plants and Morgan's fruit flies were chromosomal mutations, the result of flipping or moving chunks of DNA (inversions and translocations) to another chromosome. A region of a chromosome can also be deleted or duplicated, perhaps leading to a 'copy number variation' that reduces or raises the number of genes and, in turn, dampens or boosts protein levels.

Changing a single letter of DNA results in a 'point mutation'. Although small, such mutations can have a large effect if they hit a gene. The genetic code dictates that genes are read in frames of three-letter words, so adding or removing one or two bases creates a 'frameshift' mutation that changes instructions into gibberish, in the same way that removing the first letter in 'bat ate bug' makes 'ata teb ug-'. When one base is substituted for another, it can also introduce a 'nonsense' mutation that translates to 'stop' in the genetic code, or change a 'stop' into an amino acid, resulting in an abnormally elongated protein.

> Gene mutations form the chief basis of organic evolution, and therefore of most of the complexities of living things.
> Hermann Müller

Single-letter substitutions can also change the translation of a genetic word to produce the wrong amino acid. One such 'missense' mutation occurs in a human gene for haemoglobin: it substitutes a water-hating molecule for a water-loving one and causes proteins to clump together. This causes normally disc-shaped red blood cells to bend into a crescent – sickle cell anaemia. This condition is harmful if you have two copies of the mutant gene, because deformed cells are

How common are new mutations? By comparing DNA sequences of family members, geneticists have found that the point mutation rate is 0.000000012 per base per generation, or one in every 83 million letters. Given that the human genome is over 3 billion letters long, that means you inherited about 40 new mutations on top of the unique combination of chromosomes from your parents. Mutations can only be inherited if they occur in a sperm or egg cell – the 'germline' – and not in body cells of somatic tissue. (This is why somatic hypermutation, the DNA changes in antibody genes that allow acquired immunity, are not passed from parent to child.) In 1947, British biologist J.B.S. Haldane proposed that males contribute more mutations to evolution than females, known as male-biased mutation. There are three to four times as many mutations in the male germline because sperm cells continually divide in men, introducing mutations during DNA replication.

less able to carry oxygen and clog blood vessels, potentially choking tissues and leading to a heart attack or stroke. But the disease can also be helpful in the right environment: carrying only one mutant gene protects people from malaria, a parasite that has trouble infecting sickle-shaped cells, so the mutation is favoured by natural selection in populations with endemic malaria.

Mutagens

Spontaneous mutations can be triggered by physical, chemical or biological sources. One of the most common is heat. Another is ultraviolet light, which changes letters from C to T and makes small bulges in the double helix. DNA repair machinery in epidermal cells can fix these mutations, but people with xeroderma pigmentosum, a rare condition of defective repair genes, have a 100 per cent chance of developing skin cancers unless they stay indoors. Natural mutagens, like reactive oxygen species ('free radicals'), are produced and mopped up by normal cellular processes to prevent DNA damage. Substances that trigger cancer (carcinogens) are the most well-known chemical mutagens, and the first artificial mutagen to be discovered was mustard gas, the World War I chemical weapon discovered to have mutagenic properties by geneticists Charlotte Auerbach and J.M. Robson in 1942.

For evolution, the most significant mutations are created by organisms and genetic parasites. The movement of viruses and transposable elements around a genome can shuttle DNA to new locations, for instance. Mutations are also introduced via errors during DNA replication. Spontaneous mutations are essential to evolution but can harm individuals, which is why cells play it safe by using DNA repair to fix changes.

The condensed idea
DNA changes are the source of variation and disease

10 The double helix

All cells carry biological information in DNA (deoxyribose nucleic acid), a molecule perfectly suited to its dual roles in storing and transmitting genetic instructions. This is ultimately down to the fact that DNA forms paired strands in a double helix – a structure that is easy to copy, and easy to fix.

The double helix is the definitive icon of biology. Like the atom, its beautifully simple structure is so recognizable that it symbolizes a whole branch of science. But while DNA's twisted ladder may be visually appealing, its chemical composition is far less interesting, which helps explain why the genetic material was overlooked for over 75 years.

Swiss physician Friedrich Miesler switched to studying biochemistry with the modest aim of discovering life's building blocks. He first focused on proteins, which were vital to cell function and abundant in the cytoplasm. But in 1869, in tests where acid was added to white blood cells extracted from pus, a phosphorous-rich compound precipitated from the solution. Realizing it came from the nucleus, Miesler called it 'nuclein'. Others started studying the compound, including another German, Albrecht Kossel, who showed that it had ribose sugars and four bases. Nuclein was later renamed deoxyribose nucleic acid – DNA.

Chromosomes, a mixture of protein and nucleic acid, were shown to carry genes in 1910. At the time, the majority of biologists believed that protein was the genetic material, mainly because its building blocks are more varied and complex: 20 amino acids versus four bases. DNA was boring, and simply did not excite scientists.

Solving the structure

By the mid-20th century, some molecular biologists had become convinced DNA was important, including the Cambridge-based American-British duo of James Watson and Francis Crick. Meanwhile, at Maurice Wilkins's lab at King's College London, Rosalind Franklin was shooting X-rays through crystals containing DNA molecules. In 1951, images from the pattern of diffracted rays made Franklin think DNA was a spiral. She changed her mind, but when Wilkins showed

Franklin's images to Watson, they inspired him. In early 1953, Linus Pauling, the American chemist who won a Nobel Prize for chemical bonds, proposed a triple helix. Watson and Crick's boss, Lawrence Bragg, had a longstanding rivalry with Pauling, and the search for a solution to DNA's structure soon became a race.

Watson and Crick combined their knowledge of crystallography and chemistry to build plastic models of DNA by trial-and-error. Crick realized that two strands could be anti-parallel, with spirals twisting in opposite directions, so each helix was a phosphate-sugar backbone with bases facing inwards like the teeth of a zipper. The next step was to fit DNA's bases between the two strands. In 1950, Austrian biochemist Erwin Chargaff had discovered that DNA samples contain equal amounts of the bases adenine and thymine (A and T), while the cytosine to guanine (C to G) ratio is also equal. This finding, Chargaff's 'parity rule', explains how two helices stay together: each base on one strand pairs with its complementary partner on the other strand, A with T, C with G. The structure of the double helix had been solved.

> We often worried that the correct structure might be dull ... The finding of the double helix thus brought us not only joy but great relief.
> James Watson

DNA replication

Watson and Crick's 1953 research paper, 'Molecular structure of nucleic acids', includes one of the greatest understatements in scientific history: 'It has not escaped our notice that the specific pairing we have postulated immediately suggests a possible copying mechanism for the genetic material.' Their second study, published a month later, 'Genetical implications of the structure of deoxyribonucleic acid', elaborated on the mechanism: the two strands separate during replication, each of which is used as a template to make a duplicate of its former partner. This hypothesis was proven true in 1958 by Americans Matthew Meselson and Franklin Stahl, who studied bacteria to show that each new double helix consists of one strand from a previous molecule, which meant that DNA replication is 'semi-conservative'.

Molecular biologists now know that the process of replication is complex and involves many enzymes, including DNA polymerase.

Solid proof that DNA – and not protein – is the molecule that carries genes came from Oswald Avery, Colin MacLeod and Maclyn McCarty, in 1944. Back in 1928, British bacteriologist Frederick Griffith had shown that after a deadly strain of pneumococcus was destroyed by heat, its remains could transform non-virulent microbes into a new strain capable of killing mice. Griffith believed his bacteria had changed by absorbing a 'transforming principle'. The Canadian-American team of Avery, MacLeod and McCarty separated the deadly strain's remnants into fractions and, through a process of elimination using enzymes to digest specific molecules, proved that the 'transforming principle' was DNA. In 1952, American geneticists Alfred Hershey and Martha Chase confirmed DNA's role in heredity by showing that when a bacterium is infected by the bacteriophage T2 virus, radioactively labelled viral proteins did not enter the cell, but viral DNA did. Biologists have since found that while all cells use DNA as the genetic material, viruses can use other structural arrangements of nucleic acid, including double-stranded RNA and single-stranded DNA.

However, the double helix is also universal, highlighting why its structure is so well suited to transmitting hereditary information: complementary strands make it easy to copy nature's instructions. This explains why, on 28 February 1953, Crick ran into his local pub in Cambridge and declared that he and Watson had discovered 'the secret of life'.

DNA repair

Although DNA is a relatively stable molecule, its suitability for genetic storage is not chemical, but due to its ability to fix potentially harmful mutations. In 1949, American microbiologists Albert Kelner and Renato Dulbecco independently described the damaging effects of ultraviolet light. Kelner was studying *Streptomyces* fungi, while Dulbecco was looking at virus-infected bacteria. Both noticed that organisms would recover from UV exposure after being placed in visible light, a process called 'enzymatic photoreactivation' that fixes small bulges in the double helix which would otherwise obstruct a cell's ability to read DNA.

Left uncorrected, mutations accumulate rapidly as cells divide and DNA replicates, increasing the likelihood of harmful changes to a gene. Thousands of random DNA lesions are triggered every day, but few become permanent thanks to various RNA repair processes. Base excision repair fixes specific letters, for example, as DNA glycosylases spot changes by flipping bases like an overeager dentist checking for rotten teeth. Nucleotide excision repair, meanwhile, fixes several letters at a time.

Damage to both strands of DNA is a potential disaster for cells, so those mutations are fixed fast. Non-homologous end-joining files off a few bases, then sticks the broken ends together – a quick-and-dirty solution that leaves a mutation. On the other hand, recombination provides a more accurate method, homologous end-joining, as DNA from a matching chromosome becomes a template for repair. This illustrates a key benefit of the double helix: each strand serves as a backup copy for each other, so if one is damaged, the other strand can be used to recover any missing genetic data.

The condensed idea
DNA's structure is built for replication and repair

11 The genetic code

Nature's code lets cells translate the instructions encoded in DNA into the language of proteins. Although the phrase is often used to describe our genetic makeup or DNA, the genetic code is an actual code – a set of rules for how information is converted from one form to another.

Genes and proteins are written in different chemical alphabets. DNA uses four 'letters': the bases adenine (A), cytosine (C), guanine (G), and thymine (T). The protein alphabet, meanwhile, has 20 amino acids. By the mid-20th century, biochemists knew that nucleic acids and proteins are formed from chains of building blocks, and in 1902, Franz Hofmeister and Emil Fischer independently proved that proteins contain amino acids.

After James Watson and Francis Crick showed that both strands of DNA are a sequence of nucleotides, each carrying one of the four bases, Crick proposed that the order of letters spells-out a protein's amino acids. This 'sequence hypothesis' is made possible by the genetic code, the rules that allow the language of genes to be translated into the vocabulary of proteins.

Cracking the code

The race to crack the genetic code started as soon as Watson and Crick revealed the double helix in 1953. Scientists initially had no idea how life's code words were written, only that DNA has four bases. If each word was two letters long, there would only be 16 (4x4) combinations – not enough for the 20 amino acids used in proteins. They guessed that the code used three-letter words, which gives 64 (4x4x4) triplets, a theory proven correct by a British team in 1961.

Knowing that proteins are encoded in a 'reading frame' of non-overlapping triplets, the next step was to decipher the meaning of each three-letter code word. In 1954, the physicist George Gamow founded the 'RNA Tie Club', a society of 20 geniuses interested in cracking the genetic code, each of whom wore an amino-acid-themed tie. Although Watson, Crick and several other Nobel prizewinners were part of the club, they did not manage to decipher the code. The first to crack a key 'word' were biochemists Marshall Nirenberg and

Heinrich Matthaei, who revealed that the triplet 'UUU' encodes the amino acid phenylalanine. By 1966, Nirenberg's lab and teams led by Severo Ochoa and Har Gobind Khorana had eventually deciphered all 64 triplets.

Transcription

So why are proteins read from RNA, not DNA? The answer comes down to how genes are controlled: the process of transcription. In 1961, François Jacob and Jacques Monod figured this out by studying the 'lac operon' in *E. coli*, a cluster of three genes for lactose metabolism. The French molecular biologists showed that the genes are controlled by DNA sequences upstream from the operon – genetic switches that can be flipped on or off by the presence of sugars like lactose. Because genes are on DNA and proteins are made in a cell's cytoplasm, Jacob and Monod proposed that the protein-making instructions are carried by an intermediate molecule: ribose nucleic acid. RNA is often a single strand and uses the base uracil (U) instead of thymine (T), but in terms of carrying genetic information, RNA and DNA are identical.

> It is one of the more striking generalizations of biochemistry ... that the twenty amino acids and the four bases are, with minor reservations, the same throughout Nature.
> Francis Crick

Transcription – reading genes to make an RNA copy – resembles DNA replication. The double helix is unzipped by cellular machinery – enzymes such as RNA polymerases – so DNA's letters can be copied to an intermediate molecule, messenger RNA (mRNA). In eukaryotes, the mRNA is then exported from the nucleus.

Translation

Based on the available chemical bonds in nucleic acids, Crick realized it was unlikely that DNA/RNA acted as a direct template for making proteins, and proposed that they would link to amino acids via small adapter molecules. In 1958, American scientists Mahlon Hoagland and Paul Zamecnik showed that radioactively labelled amino acids would become attached to RNA before being incorporated into proteins, implying that RNA transfers amino acids during protein

The central dogma

Genetic information can only flow in certain directions. This is the 'central dogma of molecular biology', sketched out in 1956 by Francis Crick: 'Once information has got into a protein it can't get out again', where information is the sequence of amino acids. James Watson, the other co-discoverer of the double helix, later caused confusion by incorrectly simplifying this concept to 'DNA makes RNA makes protein'. Crick refined the central dogma in 1970 by defining three types of transfer: general transfers that occur in all cells, such as DNA to DNA (replication), DNA to RNA (transcription), RNA to protein (translation); special transfers include RNA to DNA, a 'reverse transcription' process used by viruses made of RNA; and unique transfers convert protein information to DNA, RNA or protein. The 'back translation' from protein to DNA or RNA should be impossible because information is lost because of degeneracy in the genetic code, but there is one example of limited information transfer between proteins: prions (see chapter 24).

synthesis. Then in 1965, biochemist Robert Holley revealed the structure of the mysterious molecule, which resembles a clover leaf. His 'transfer RNA' (tRNA) carried the amino acid 'alanine', and proved Crick's adapter hypothesis correct.

The process of translation is performed by another cellular machine, the ribosome. As an mRNA strand is pulled through the ribosome like paper through a printer, one tRNA adapter at a time will attach itself to each three-letter word or 'codon' using matching 'anti-codon' triplet from the tip of tRNA's clover-leaf structure. The ribosome adds that tRNA's attached amino acid to a growing chain that will be folded into a 3D protein.

Degeneracy

The genetic code's triplet nature gives it an important feature: it is 'degenerate'. Just as synonyms are words with a similar meaning, degeneracy means several codons are translated as the same amino acid. This occurs because genes use 64 codons but proteins use 20 words, so most amino acids are encoded by several triplets (three codons act as a 'stop' for translation, leaving 61). One consequence of degeneracy is the 'central dogma': information is lost in translation – the equivalent of the genetic language having words for 'cat' and 'dog', but proteins only understanding 'pet'. Degeneracy means that substituting one base for another might not change an amino acid. All codons starting 'GG' (GGA, GGC, GGG, GGU) translate to 'glycine', for example. Base substitutions that create synonymous codons are 'silent' mutations, but are not always harmless (see chapter 12).

While mitochondria and some microbes use slight variations, the vast majority of organisms use a standard genetic code. This is no accident, but a result of natural selection. In 1998, evolutionary biologists Stephen Freeland and Laurence Hurst ran computer simulations to generate random codes with different rules, then assessed the impact of mutations. Out of a million random codes, just one was better at minimizing the effect of mutations.

The condensed idea
Rules translate DNA's instructions to make proteins

12 Gene expression

An organism's characteristics are ultimately determined by its genes, which encode proteins that dictate the features of cells. The genetic instructions specify more than just the sequence of a protein – they influence biological complexity.

The invention of molecular biology techniques, such as mutagenesis in 1927, let scientists study genetic effects at the biochemical level. At the same time, researchers widened the array of model organisms beyond lab workhorses like the fruit fly *Drosophila melanogaster*. Bernard Dodge of the New York Botanical Garden, working with the bread mould *Neurospora crassa*, once visited Columbia University and told Thomas Hunt Morgan: 'It's even better than *Drosophila*'. In 1928, when Morgan moved to the California Institute of Technology (Caltech), he took the mould with him.

American geneticists George Beadle and Edward Tatum also saw potential in *Neurospora*. Beadle, who had studied fruit flies at Caltech in the 1930s, realized the mould would be ideal for observing the effects of mutation on metabolism, the biochemical reactions that keep living things alive. *Neurospora* makes its own nutrients from food, but when Beadle and Tatum exposed it to X-rays in 1941, some moulds lost the ability to produce things like vitamin B6, growing in Petri dishes only when supplied with the missing nutrient. This showed that genes work at specific points in a metabolic pathway, suggesting they make the enzymes that catalyse biochemical reactions. This is the 'one gene, one enzyme' hypothesis.

Proteins play numerous roles beyond acting as enzymes. In the 1960s, the genetic code revealed that a DNA sequence specifies the amino acids in a protein – a polypeptide chain – and so the hypothesis became 'one gene, one protein'. The steps from gene to protein were also shown: instructions in DNA are transcribed (read and copied) to RNA, then translated (decoded) to protein. But this process – gene expression – includes many other steps. DNA might be unwound from other molecules in a chromosome, for example, and a polypeptide only becomes a protein once folded into a 3-D shape. Two steps stand out because they reveal key features of genes: switches and pieces.

The genetic code is 'degenerate' because 64 three-letter codons (code words) are translated into 20 amino acids. As most amino acids are encoded by two or more codons – 'GGA' and 'GGG' both specify glycine, for example – then some changes do not alter a protein sequence, so it was once assumed that these mutations were 'silent' to an organism's features and therefore natural selection. But in the 1980s, geneticists like Richard Grantham and Toshimichi Ikemura noticed something strange – some codons were preferred over others: although both 'AAC' and 'AAT' encode asparagine, 'AAC' is more common in *E. coli* DNA. This 'codon usage bias' varied between species and matched levels of tRNA molecules, suggesting the bias improved translation. The pattern was found in everything from yeast to fruit flies – but not in mammals, even though comparing genes in different species showed that certain DNA changes were avoided during evolution. So what was going on? While certain codons may not be important in terms of a protein sequence, we now know they are needed for correct gene expression. The splicing machinery needs specific letters to identify exons, for example. Silent mutations even cause disease in humans, proving they are not so silent after all.

Genetic switches

How do cells control when proteins are made? The essentials of transcription were shown by François Jacob and Jacques Monod in 1961, based on a cluster of three genes for lactose metabolism: the 'lac operon' in *Escherichia coli*. The biologists showed that molecules now called 'transcription factors' attach themselves to DNA sequences near genes, turning them on or off. For the lac operon, switches were flipped when sugars such as lactose were present, but most genes are controlled by signals telling transcription factors to bind to DNA. Because RNA is readily broken down by enzymes, Jacob and Monod suggested that it serves as a temporary message. Like energy-saving lights in a hallway controlled by a push-button timer, RNA's transient nature means DNA switches regulate protein synthesis directly, as proteins are only made while DNA is actively copied to RNA.

Transcriptional regulation is more elaborate in eukaryotic cells. Eukaryotes have more genetic switches than bacteria (where transcription and translation occur simultaneously) and the main switch – the promoter – lies just upstream of where transcription starts. Enhancer switches, far away on a chromosome, are brought close to the promoter by transcription factors that attach to DNA and each other, like pinching ends of string to create a loop. Transcription factors prompt the RNA polymerase enzyme to transcribe DNA to a messenger RNA copy.

> Biochemists recognize the genetic material as an integral part of the systems with which they work.
>
> George Beadle

Given that an organism's features are determined by proteins, you might expect differences between individuals to come down to the protein-coding sequences in DNA. But in humans at least, this is not true: when two unrelated people are compared, their coding sequences are on average 99.9 per cent identical. So what makes us unique? In 2003, the US National Human Genome Research Institute launched the ENCODE (Encyclopaedia of DNA Elements) project to identify all functional sequences in the genome, which revealed that genetic switches are responsible for a lot of the variation among individuals. Changing DNA letters affects a transcription factor's ability to attach to specific sequences on a switch, which in turn affects how cells read DNA. Genetic activity is therefore not an on/off toggle, but has a 'dimmer switch'.

Genes in pieces

If genes determine features, it would seem logical for complex species to have more genes. In the 1990s, this reasoning led many biologists to predict the human genome would have 100,000 genes. But when the first draft of our complete DNA sequence was published in 2001, it was revealed to contain only 30,000, and recent estimates put it at a mere 20,000 genes – about the same as a nematode worm. Yet the human body has 37 trillion cells of 200 different kinds, while nematodes are 1 millimetre long and have 1000 cells.

The secret to complexity lies in split genes. For most of the 20th century, Morgan's analogy of 'beads-on-a-string' shaped how scientists saw a gene, as a distinct bit of DNA on a chromosome. But while studying adenovirus in 1977, Richard Roberts and Phillip Sharp discovered that when RNA sequences were mapped onto its matching DNA, the DNA's length was much longer than the RNA, revealing that genes consist of pieces. Studies of non-viral genes such as haemoglobin confirmed that cells also have split genes. The beads-on-a-string model may not work for genes on a chromosome, but it may apply to protein-coding sequences, where genes are split into 'exons' interrupted by long stretches of 'introns'. The average human gene contains 10 introns, but the longest (encoding the muscle protein titin) has 363 exons. As introns do not encode genetic instructions, they are removed when DNA is transcribed into pre-mRNA (precursor messenger RNA). This 'RNA splicing' process is carried out by a 'spliceosome machine' containing several catalytic RNAs and hundreds of proteins. The spliceosome makes a loop between two exons that may contain a single intron or multiple segments, snips it out, then joins the exons. The loop might be a single intron or multiple segments.

Like a movie editor splicing scenes from a film reel, cells can cut-and-paste exons to produce different combinations of mRNA – multiple proteins from one gene. This 'alternative splicing' generates a staggering diversity of proteins. The Dscam gene in fruit flies, for example, has 95 exons and is capable of making over 38,000 proteins. This helps explain how organisms can be more complex without more genes: about 20 per cent of genes in the nematode worm are alternatively spliced, whereas over 90 per cent of humans genes encode multiple proteins.

The condensed idea
DNA directs the protein-making process

13 Protein folding

Proteins do almost all the hard work in living organisms, from catalysing cellular metabolism to connecting body tissues. Such functions require chains of amino acids to fold into three-dimensional shapes, but understanding how this happens has proven to be the biggest challenge in molecular biology since the race to crack the genetic code.

American chemist Linus Pauling won two Nobel prizes, and might have become the only person with three, had he beaten Watson and Crick in the race to solve the structure of DNA. His first Nobel, in Chemistry (the second was for Peace), was for work on the quantum nature of chemical bonds and the structure of complex substances, such as proteins. While a visiting professor at Oxford University in 1948, Pauling had caught a cold and ended up in bed, where he quickly got bored of detective novels and started making molecular structures out of paper. Hours later, the creative genius had constructed a spiral held together by hydrogen bonds at regular intervals along the chain. Upon returning to the California Institute of Technology, he worked with X-ray crystallographer Robert Corey and physicist Herman Branson to confirm that his paper model was correct, and in 1951, he revealed his discovery – the alpha-helix.

Structures

The primary structure of a protein is its amino acid sequence, and this 'polypeptide chain' coils, kinks or twists into a secondary structure – either an alpha-helix, a beta-sheet or turn. A protein is formed by folding into a three-dimensional shape – the tertiary structure. This may stand alone, or form part of an even more complex quaternary structure (as in haemoglobin, which consists of four subunits). Proteins can be globular, form part of membranes or produce fibres. The fibrous connective tissue that holds cells together – collagen – makes up about one third of the human body.

The first 3-D structure of a protein was presented in 1958 by British crystallographer John Kendrew, who showed off the long, winding sausage-like shape of myoglobin, the oxygen-carrying molecule in muscle tissue. As he described it: 'The arrangement seems to be almost totally lacking in the kind of regularities which one instinctively anticipates, and it is more complicated than has been predicated by any theory of protein structure.' This led biologists to ask several questions. How does an amino acid sequence encode structure? What allows proteins to fold so fast? And can structure be predicted from a sequence? Together, these are known as the protein-folding problem.

The protein code

Researchers once hoped that the mysteries of protein-folding might be solved by a code featuring simple rules, akin to the base pairing between opposite strands in DNA. But things are not so easy: the Protein Data Bank, an online repository established in 1971 and now holding 100,000 structures in atomic detail, describes hydrogen bonds, close-range van der Waals forces, preferred angles in polypeptide backbones, plus electrostatic and hydrophobic interactions between amino acids. Understanding such interactions might ultimately lead to a series of coding rules.

> If you want to have good ideas you must have many ideas. Most of them will be wrong, and what you have to learn is which ones to throw away.
> Linus Pauling

In the 1960s, American biochemist Christian Anfinsen studied a small catalytic protein called a ribonuclease. Like all enzymes, its 'active site' contains atoms that interact with a specific molecule to catalyse a chemical reaction. As reactants bind onto the active site, an enzyme changes conformation (shape) and releases products. In 1961, after altering a solution so enzymes adopted a denatured (non-functional) shape, Anfinsen proved that a protein can be refolded back to its native conformation. This led him to suggest that all the information needed for producing a protein is encoded within the polypeptide. As stated in 1973: 'the native conformation is determined by the totality of inter-atomic interactions and hence by the amino acid sequence'.

Anfinsen's dogma, as this rule is known, was originally called the 'thermodynamic hypothesis'. Basically, while folding, a protein is driven towards a state with the lowest free energy, creating thermodynamically stable molecules. Scientists imagine this pathway

Proteins have diverse functions in organisms, but arguably their most vital role is as enzymes – catalysts that drive the metabolic reactions behind life. Metabolism includes thousands of biochemical processes. Anabolic reactions involve construction, such as fatty acid synthesis from sugar. Catabolic reactions involve breakdowns, like digestion of starch into sugars. Mutations in genes that encode enzymes cause disease, as first shown by British physician Archibald Garrod. In 1908, Garrod suggested that alkaptonuria, a rare inherited disorder whose symptoms include black urine and joint pain in middle age, resulted from the inability to break a chemical bond in alkapton or 'homogentisic acid', itself due to a problem with a specific enzyme. Garrod later classed congenital disorders like alkaptonuria as 'inborn errors of metabolism', connecting inheritance with proteins for the first time.

as an 'energy landscape' shaped like a funnel: the polypeptide has room to explore a different conformation at the top, but fewer options as the funnel gets narrower. This helps visualize a folding pathway, but not the natural process.

Fast folding

In 1969, American molecular biologist Cyrus Levinthal gave a talk entitled 'How to Fold Graciously' about how temperature denatures and renatures enzymes, providing a back-of-the-envelope calculation for the number of possible combinations in a theoretical protein. This highlighted the fact that a polypeptide chain can theoretically adopt an astronomical number of structures, and yet it finds the right one almost spontaneously, sometimes within microseconds. One resolution to 'Levinthal's paradox' is that local secondary structures fold first – maybe even while the chain is being made – and global folding happens later, vastly reducing the number of theoretical variations. Since the 1980s, biologists have also known that cells contain 'molecular chaperones' to help guide folding and refolding.

Predicting structures

Since 1994, the Critical Assessment of protein Structure Prediction (CASP) has tested computer software tools that predict 3D shapes against what has been visualized using experimental techniques like X-ray crystallography or cryo-electron microscopy. In 2020, the CASP competition saw a breakthrough: software whose predictions from amino-acid sequences alone were (on average) at least as accurate as the real structures determined by experiments. The winning entry was developed by artificial intelligence (AI) firm DeepMind, part of Google's parent company Alphabet, and is called AlphaFold.

Despite its name, however, AlphaFold does not actually predict a structure through folding, it builds the shape based on a dataset of known interactions in repositories such as the Protein Data Bank (which stores over 200,000 structures). This means AlphaFold cannot predict how a protein's structure might be affected by mutations that alter its amino acids or by a molecule (like a drug) that binds to the protein and prompts structural changes, limiting the software's potential to study molecular evolution or the discovery of new drugs. That said, the breakthrough suggests it may not be long before AI tools can fold, too, and reveal more about how proteins work.

The condensed idea
Predicting structure to understand the function of biomolecules

14 Junk DNA

The genome is an organism's entire complement of DNA, a full set of chromosomes carrying all its genes. In many species, including humans, only a small part of the genome consists of protein-coding genes. The rest is non-coding, sometimes referred to as the genome's mysterious 'dark matter'. But is this junk?

Your genome is five times smaller than an onion genome – a fascinating fact that makes no sense if you think that a complex organism should have more DNA. In fact, when its genome length is compared to more complex organisms, the vegetable often wins – a comparison Canadian geneticist T Ryan Gregory calls 'the onion test'. This is related to the 'C-value paradox' highlighted by Harvard physician C.A. Thomas Jr in 1971: the size of an organism's genome does not reflect its biological complexity.

Part of the reason why DNA and complexity are not related is that genomes contain varying amounts of useless rubbish, or 'junk DNA'. This term was already being used in the 1960s, but was made famous by Japanese geneticist Susumu Ohno in 1972. Ohno argued that if every letter in a DNA sequence was useful then the burden of harmful mutations would be unbearable. Instead, he said: 'The Earth is strewn with fossil remains of extinct species; is it a wonder that our genome too is filled with the remains of extinct genes?'

Selfish DNA

Back in 1953, American geneticist Barbara McClintock reported that sections of chromosomes in maize plants would move around during cell division. McClintock's discovery of 'jumping genes' was largely ignored until the late 1960s, when they were found in other organisms. In 1980, several prominent scientists suggested that most DNA in the genome of a complex organism is 'selfish'. Jumping genes (now called 'transposable elements') are one example. Some fragments of this parasitic, selfish DNA can still move around a genome by copy-and-paste or cut-and-paste, while others lose that ability, remaining relatively harmless to their host.

The draft sequence of the human genome, published in 2001, revealed that transposable elements constitute 45 per cent of our

Copy number variation

We are not 99.9 per cent identical. The figure is based on aligning DNA from two unrelated individuals side-by-side and counting small differences like single-letter changes in the sequence. In 2004, Charles Lee and Michael Wigler independently discovered that the human genome contains significant copy number variation, or CNV: large sections of DNA are deleted or duplicated. This means that instead of a few 'spelling mistakes', the difference in the book of life between two people is like having missing or extra pages or chapters. This can have no impact, change features or cause disease. Some CNVs contain genes. For example, while some people inherit one copy of the amylase gene – encoding the enzyme for digesting starch – from each parent, others can have up to 16 copies. In 2015, Stephen Scherer compiled DNA sequences from healthy individuals from diverse ethnic backgrounds to annotate a map of the human genome with CNVs, and found that about 100 genes can be completely deleted with no ill effects. Rather than a 0.1 per cent difference between people, the CNV map shows that 5–10 per cent of the genome consists of CNVs.

DNA, but mutations can make those ancient sequences unrecognizable over time: recent estimates indicate that transposable elements make up between half and two-thirds of our genome. In maize plants, the figure is closer to 90 per cent. This difference goes a long way towards explaining the onion test.

Non-coding DNA

The human genome is 3.2 billion base pairs (letters) long. Printed onto pages in barely readable text, it would fill encyclopaedia-sized tomes on a bookshelf from floor to ceiling. We have around 20,000 genes, DNA sequences that encode proteins, but they only make up 1.2 per cent of the total genome; the remaining 98.8 per cent is referred to as non-coding DNA. However, non-coding DNA and junk

DNA are not the same thing. Coding determines whether DNA makes proteins, but whether DNA is 'junk' is a question of whether it is useful to the organism carrying it. Think of the genome as a vast junkyard. Employees have driven to work so you know their parked cars are functional, or 'coding'. What about all the 'non-coding' vehicles in the junkyard? Some is scrap metal, others can be salvaged, a few might even be functional cars parked in the junkyard without you noticing. The only way to tell is to examine every vehicle. The flip side of this (a logical error made by creationists) is to think that if you find one useful item, the whole junkyard must be useful.

Functional DNA

Along with its protein-coding genes, DNA also carries RNA-specifying genes. Well-known RNA genes produce tRNA molecules used to translate the genetic code, but many more are known to be vital because mutations in the RNA gene causes disease. DNA also has control elements such as genetic switches that regulate gene expression. So how much non-coding DNA is useful? Scientists do not agree on a figure.

One approach, used by geneticists, is to align the genomes from two species – such as human and another mammal – side-by-side to calculate how much of the DNA sequence has been conserved over time. This gives 5–9 per cent. Another approach is to measure how DNA interacts with other molecules. The largest effort to identify functional DNA in this way was the ENCODE (Encyclopaedia of DNA Elements) project. In 2012, the ENCODE team concluded that 80 per cent of the genome had a 'biological function'. Many biologists criticized this claim, as it used a loose definition of 'function' that is closer to 'biological activity' than whether the DNA is useful for an organism.

Harmless DNA

Misunderstanding by the public, journalists and even scientists is ultimately due to inappropriate use of synonyms. Junk DNA is 'junk' – not 'trash' or 'waste'. This was articulated in 1998 by South African geneticist Sydney Brenner: 'There is the rubbish we keep, which is junk, and the rubbish we throw away, which is garbage. The excess DNA in our genomes is junk, and it is there because it is harmless'.

In 2015, evolutionary biologist Dan Graur extended Brenner's idea, classifying DNA according to its 'selected-effect' function – whether the gain or loss of DNA might have an effect on an organism's fitness (ability to survive and reproduce) and so be noticed by natural selection. This produces four DNA classes: 'Literal DNA' carries information in its order of

> People feel very uncomfortable with the idea that a lot of their DNA is of no value.
> Sydney Brenner

letters (so protein-coding genes, RNA genes and control elements are in this class); 'Indifferent DNA' is functional by its presence (analogous to how spaces between words make text readable); 'Junk DNA' neither helps nor hinders fitness; and 'Garbage DNA' is harmful.

Why do organisms not remove junk DNA? Because it is has little to no effect on fitness. If it went from harmless to harmful, natural selection might remove it. As Sydney Brenner explained: 'Were the extra DNA to become disadvantageous, it would become subject to selection, just as junk that takes up too much space, or is beginning to smell, is instantly converted to garbage by one's wife, that excellent Darwinian instrument.'

The condensed idea
Genomes contain
harmless rubbish

15 Epigenetics

Although most biological instructions are encoded within DNA sequences, some are carried by chemical tags added to the genetic material and its proteins. These epigenetic marks are a record of past environments and experiences, and reveal how characteristics acquired during your lifetime can be inherited.

In 1809, French naturalist Jean-Baptiste Lamarck proposed one of the first theories of evolution, suggesting that change in the environment drives evolution of species. So far, so good – but he also claimed that body parts could be strengthened through continued use, and neglect would cause them to deteriorate, and that this improvement or deterioration during a lifetime was passed from parent to offspring – so-called inheritance of acquired characteristics. Lamarck's theory was proven wrong in an 1889 experiment by August Weismann: the German biologist cut the tails off more than 900 mice over five generations, and found that offspring were still born intact. Twentieth-century genetics showed that biological instructions are stored in DNA, but the discovery of other mechanisms of inheritance has revived Lamarck's ideas.

Inherited instructions

In 1942, British biologist Conrad Hal Waddington suggested that the then-mysterious mechanics of development were controlled by 'epigenetics' (Greek for 'above genetics'). Scientists agree that this process involves a mother cell transmitting instructions to daughter cells during division, but there is no consensus definition. The most common was proposed in 1996 by American geneticist Arthur Riggs, who defined epigenetics as 'heritable changes in gene function that cannot be explained by changes in DNA sequence'.

Early insights into epigenetics came from differences between sex chromosomes in mammals. While studying female rats in 1959, Japanese geneticist Susumu Ohno observed that one of their two X chromosomes looks compact and condensed, suggesting it is not used by cells. In 1961, British geneticist Mary Lyon suggested this could explain colour patterns in mouse fur, which are determined by genes linked to the X chromosome. Lyon proposed that the action of genes on one X chromosome is blocked – this 'X inactivation' or 'lyonization' helps explain why XX females do not produce a double dosage of X-linked proteins compared to XY males.

Cellular software

Cells are like computer hardware, DNA is the operating system, and epigenetics provides the software. Most epigenetic programming consists of chemical modifications or 'epigenetic marks' that hide DNA sequences from the cell's gene-reading machinery. As independently suggested by Arthur Riggs and Robin Holliday in 1975, one mechanism that programs a cell involves adding methyl groups to DNA, deactivating genes.

DNA methylation acts like a chemical blanket that muffles gene activity. Other epigenetic programmes hide DNA sequences without modifying genetic material itself. For example, X inactivation uses non-coding RNA molecules of 'XIST' to cover the chromosome. Another mechanism modifies giant proteins called histones: DNA strands are coiled around histones like thread wrapped around multiple bobbins, collectively called chromatin. The epigenetic marks on histones cause DNA to wind or unwind, so chromatin is 'closed' or 'open' and genes can be read. Finally, transcription factors are proteins that determine a cell's features and behaviour by binding to DNA control switches and turning genes on and off. Programming with transcription factors is closer to Conrad Hal Waddington's definition of 'epigenetics' during development.

> Whatever difficulties there may be in the discovery of new truths in nature, there are still greater difficulties in getting them recognized.
> Jean-Baptiste Lamarck

Epigenetic marks can be wiped from the genome through a process of reprogramming, leaving an embryo in a blank slate where its stem cells can form any tissue. In mammals, marks are erased in two waves: after fertilization and during embryogenesis. Post-fertilization reprogramming wipes most marks except those added to sperm or egg in genomic imprinting – akin to deleting all of a computer's programs, then installing some essential apps. The second wave, in the embryo, is a thorough erasure process that reinstalls the cell's operating system, returning the machine to factory settings.

Environmental exposure

Pregnant mothers avoid toxins such as alcohol and certain foods to avoid harming their baby in utero, while smoking by young fathers increases the body weight of sons. Exposure to factors such as hormones and stress affect the health of their future children – not because they trigger genetic mutations, but because they add chemical marks to an embryo's DNA: 'epi-mutations'. Parental effects even stretch over generations, revealing that 'you are what your grandparents ate'. During the Dutch famine of 1944, pregnant women exposed to poor nutrition produced children with poor glucose tolerance, while their grandchildren were born with extra body fat and developed poor general health in later life. Since 2001, Lars Olov

Genomic imprinting

Before you were born, your parents left their mark on your DNA, in the form of epigenetic imprints that deactivated parts of your genome. One example is insulin-like growth factor 2 (IGF2), a hormone that helps produce large children: fathers add methyl groups to the gene encoding its matching IGF2 receptor in sperm DNA, leaving IGF2 itself untouched, while mothers add DNA methylation to the IGF2 gene in eggs and leave the receptor active. This 'genomic imprinting' has evolved due to the conflict of interest between males and females over how limited resources such as nutrients should be spent on growing offspring: a dad wants to blow everything on new progeny, while a mum would prefer to allocate resources equally among her children, saving some for future kids. In mammals, the custody battle over DNA has left each sex in sole control of about 100 genes, which now need to be imprinted with epigenetic marks for normal embryonic development. In humans, not imprinting the maternal copy of IGF2 results in a double dose of proteins, leading to Beckwith-Wiedemann syndrome, a bigger baby with health problems that must be born by caesarean section.

Bygren and Marcus Pembrey have used data from Överkalix in northern Sweden to compare longevity and causes of death with historical records of harvests and food supply. They found that mortality in men is influenced by their paternal grandfather's nutrition in mid-childhood, while mortality in women is affected by their paternal grandmothers's diet, suggesting a link to epigenetic marks on sex chromosomes.

Soft inheritance

Long-term epigenetic programming in mammals is limited because virtually all chemical marks are wiped during embryo reprogramming. Thus, parental effects caused by environmental factors only last one or two generations – so-called 'intergenerational' inheritance. Longer-lasting 'transgenerational' inheritance is rare in animals, with nematode worms being a rare exception: epigenetic imprinting of an attractive smell can be transmitted over 40 generations. Permanent changes to DNA can be considered 'hard inheritance', while temporary epigenetic programming is 'soft inheritance'.

Soft inheritance is relatively common in flowering plants, because there is little reprogramming and an embryo is mostly formed from its mother's body cells, rather than a reprogrammed egg. Epigenetic marks have been transmitted over hundreds of years, too: in 1744, Swedish naturalist Carl Linnaeus described a 'monster' snapdragon flower with radial rather than two-sided symmetry, and in 1999 plant biologist Enrico Coen discovered that this is caused by an epi-mutation. So it seems that Lamarck was not completely wrong about inheritance after all.

The condensed idea
Not all information is inherited in DNA sequences

16 The phenotype

The genotype, an individual's combination of genetic variants, controls every feature from invisible biochemistry to body size. But the physical manifestation of the genotype – the phenotype – is not simply determined by DNA, but also by complex interactions between genes and the environment.

The relationship between our genes and our internal and externally visible characteristics is often oversimplified. We often loosely talk about the gene 'for' a feature like eye colour, even though in reality, most traits are actually controlled by multiple genes. Each gene also has a genotype – a combination of variants or alleles – and you can inherit different versions from your parents. The relationship between a pair of alleles usually blocks the expression of one copy, so that its influence is not displayed in the resulting phenotype. But there is also the influence of the environment to consider.

Individual variation

Natural selection cannot 'see' a genotype, only its effects on the phenotype. In 1859, Darwin called the phenotypic diversity in a population 'individual variability' but it was only after his theory of evolution was combined with genetics that different phenotypes were studied by the various genotypes, or 'genetic polymorphism'.

In 1963, however, biologist Ernst Mayr said that populations can contain several phenotypes, 'the differences between which are not the result of genetic differences'. Mayr called this variation 'polyphenism'. Polyphenism is caused by environmental factors. The contribution of genes (G) and the environment (E) is represented in an equation for the variance (V) in a given phenotype (P) within a population:

$$V_P = V_G + V_E + V_{GXE}$$

For a feature such as the seed colour of pea plants, phenotype diversity depends mostly on gene variance (V_G). If the environment (V_E) plays a part, you get polyphenism. And if there is strong gene-environment interaction (V_{GXE}), various phenotypes occur.

Flexible phenotypes

Certain features are easily moulded by the environment – the phenotype is flexible or 'plastic' because its genotype reacts to a range of external conditions. In some species, sex is determined by the surroundings. For reptiles such as crocodiles and turtles, for example, this 'environmental sex determination' is dictated by temperature, an adaptation that tweaks sex ratios to improve chances of mating. Plasticity can evolve to match predictable environmental changes, too – as in the butterfly *Bicyclus anynana*, which cycles between two phenotypes over alternating generations, so adults are adapted to either the wet or dry season. In unpredictable environments, plasticity can help organisms exploit available resources: tadpoles of the Mexican spadefoot toad have two phenotypes with different jaws, digestive systems and food preferences.

An individual's plastic phenotype is most commonly determined during development, and often influenced by the population. Ants, bees and termites have workers and queens with different roles in insect society, and an individual is assigned a caste according to nutrition at the larval stage. Phenotypic plasticity can also dictate the fate of individuals over several generations, as shown in the locust *Schistocerca gregaria*, which has two adult phenotypes: a solitary, sedentary insect with short wings; and a gregarious, migratory form with long wings. The initial cue to follow a phenotypic path is crowding among young insects, the frequency with which their hind legs are touched controls hormones that trigger metamorphosis into one of the two phenotypes. The migratory form persists for several generations after the original leg-rub stimulus through another cue – chemicals in the foam that surrounds eggs. This does not change the DNA, so the phenotype is inherited through epigenetics.

> Between genotype and phenotype, and connecting them to each other, there lies a whole complex of developmental processes.
> Conrad Hal Waddington

Phenotypic plasticity can be useful in an individual's lifetime for 'physiological adaptation'. Mammals grow thicker fur in winter, for example, and adaptive immunity helps them avoid reinfection by pathogens. Human mountaineers adapt to a thinner atmosphere

The physical manifestation of genes is usually interpreted to be biological characteristics. But Richard Dawkins, author of *The Selfish Gene*, believes that the phenotype can also apply to a feature beyond an organism's body, in their environment. In his 1982 book, *The Extended Phenotype*, Dawkins states that because the ability to build structures like a bird's nest or beaver dam is determined by genes (for, say, dexterity) and influences an individual's fitness, natural selection can also act upon these 'phenotypes'. Just as survival of the fittest picks between the best genes through their effect on biological features such as beauty or strength, a beaver with good 'dam-building genes' is more likely to survive. Phenotypes could also extend into other organisms: genes that help a parasite spread by changing its host's behaviour can be favoured by selection. Like the gene-centred view of evolution or 'selfish gene', the 'extended phenotype' is not a theory, but a way of looking at life from a certain perspective. While *The Selfish Gene* popularized the work of other scientists, Dawkins's extended phenotype concept is his own contribution to evolutionary biology.

through acclimatization, as their red blood cells alter to carry extra oxygen. Flexible connections between nerve cells – synaptic plasticity – underlie learning, memory and behavioural adaptation.

Environment and evolution

In 1896, American psychologist James Baldwin suggested that an individual's ability to learn new behaviours leads to phenotypes that are sensitive to environmental factors – the 'Baldwin effect'. Then in 1942, British biologist Conrad Hal Waddington proposed that sensitivity can be reduced through 'canalization', buffering a biological feature against environmental influence during development. Waddington also pointed out that an opposite effect allows environmental factors to affect inherited characteristics, or epigenetics.

How does phenotypic plasticity evolve? One theory involves genetic accommodation: over time, the genotype adjusts to accommodate environmental changes, strengthening or weakening its control over the phenotype. A new phenotype is created through genetic mutation or sensitivity to environmental change, and once natural selection can 'see' the effect of the underlying genotype, it is exposed to survival of the fittest: if the phenotype improves an individual's fitness, those genes spread through the population.

The extended synthesis

According to the 'modern evolutionary synthesis', an environment poses problems for organisms and genes provide the solutions for adaptation. Simply put: environment proposes, genetics disposes. But a growing number of scientists argue that, just as genetics was added to natural selection to produce the modern synthesis, biology's central theory should also include the interactions between genes and environment – phenomena such as epigenetics and phenotypic plasticity – to create an 'extended evolutionary synthesis'.

This view has been promoted in some high-profile books, but not all experts agree that evolutionary theory needs updating. A counter-argument is that genetic mutations create most variation among individuals, and the phenotype is the ultimate result of a genotype – genes drive evolution. On the other hand, natural selection cannot see genotypes directly, only their effects on the phenotype, so you could say that genes are being dragged along for the ride. Are genes 'leaders' or 'followers' in evolution? Like 'nature versus nurture', the two options may not be mutually exclusive.

The condensed idea
Physical features are determined by both genes and environment

17 Endosymbiosis

O nce upon a time, a free-living bacterium found itself inside a host cell, and the pair began a relationship that evolved into marriage. The bacterium then lost its independent lifestyle and most of its belongings. This fantastic scenario has happened several times during the origin of the eukaryotic cell.

The eukaryotic cell is a complex structure – DNA is enclosed by the nucleus, so cell division requires the multi-step process of mitosis. It contains mitochondria dedicated to respiration, and plastids in photosynthetic organisms. Until the mid-20th century, it seemed unlikely these last two structures evolved from bacteria – then along came Lynn Margulis to champion the theory. In 1966, the American biologist collected the evidence from physiology and biochemistry that showed similarities between bacteria and the structures inside eukaryotic cells. After being rejected by over a dozen scientific journals, her paper 'The Origin of Mitosing Cells' was accepted by *The Journal of Theoretical Biology* a year later. She received 800 requests for reprints.

The endosymbiotic theory

Whereas intracellular parasites such as viruses are unwelcome invaders, endosymbiosis (Greek for 'within' and 'living with') is the relatively peaceful coexistence of one organism inside another. In 1905, Konstantin Mereschkowski, a Russian biologist who studied lichen (a symbiosis between fungi and photosynthetic algae or bacteria) proposed that chloroplasts evolved from an endosymbiont. American biologist Ivan Wallin suggested a similar origin for mitochondria in 1923.

The idea that certain eukaryotic organelles – mitochondria and plastids – originated from other organisms was first suggested by German biologists. In an 1883 study where he coined the term 'chloroplast', botanist Andreas Schimper said that green plants 'owe their origin to the union of a colourless organism with one evenly stained with chlorophyll'. Then in 1890, Richard Altmann noticed structures he named 'bioblasts' were ubiquitous in large cells and resembled bacteria, suggesting they were 'elementary organisms' with vital functions. Bioblasts are mitochondria.

Several features support the idea that mitochondria and plastids evolved from bacteria. They have similar sizes and shapes, for example, and divide by simple binary fission, not mitosis. For decades, endosymbiosis was not taken seriously though. Views began to change in 1962, after biologists Hans Ris and Walter Plaut looked at the green alga *Chlamydomonas* under an electron microscope. Dyes that stained genetic material were visible in chloroplasts, and DNA-digesting enzymes caused the colour to disappear. Using similar techniques in 1963, Margit and Sylvan Nass showed that mysterious fibres in mitochondria contain DNA. After Lynn Margulis published her 1967 paper, she expanded upon the evidence in a 1970 book, *Origin of Eukaryotic Cells*.

Definitive proof came from genes. Before the 1980s, there were several theories for organelle origins, such as inward folding of the cell membrane or budding of the nucleus. These theories predicted that if organelles had an internal origin, their genes should be more similar to those in the nucleus than to genetic material in free-living bacteria. In 1975, American biochemists Linda Bonen and W. Ford Doolittle found that genetic sequences from the chloroplasts of the red alga *Porphyridium* resembled material from prokaryotes called cyanobacteria. Later genetic comparisons revealed that mitochondria are descended from alpha-proteobacteria.

> The eukaryotic cell is the result of the evolution of ancient symbioses.
> Lynn Margulis

Mitochondria

In most (but not all) eukaryotic cells, mitochondria generate the energy-carrying molecule ATP by using oxygen to burn carbohydrates through aerobic respiration. Because every eukaryotic branch on the tree of life has some sort of mitochondrion, different types probably descended from one bacterium and adapted to their host's environment. By comparing the genes involved in energy metabolism from various proteobacteria, Italian researchers say that the closest living relatives are 'methylotrophs' – microbes whose outer membrane resembles the folds inside mitochondria, which is where energy is generated.

According to American evolutionary biologist William Martin, the initial symbiotic relationship provided the host cell with a source of hydrogen for generating energy, after which the last eukaryotic common ancestor (LECA) ended up generating far more energy than prokaryotes can produce. Martin and British biochemist Nick Lane say this extra energy allowed eukaryotes to add genes to their genome, enabling them to build complex cells. If this hypothesis is true, gaining mitochondria was the crucial step on the path to the eukaryotic cell – before it gained its defining feature, the nucleus, about 1.5 billion years ago.

Plastids

Plastids make carbohydrates using pigments that capture energy from light. There are two types: primary plastids with two membranes (which include the famous green chloroplasts of plants), and secondary plastids, with three or four. The double membranes of primary plastids (and mitochondria) have different molecular compositions: the inner one resembles a bacterial membrane; the outer one is like the surface of a eukaryotic cell. This is consistent with a cyanobacterium being engulfed in a bubble formed from the host membrane, about 1.2 billion years ago.

Primary plastids have evolved along with their hosts into three branches in the tree of photosynthetic life: freshwater algae (Glaucophytes), red algae (Rhodophytes) and 'plants' (green algae and land plants). Because secondary plastids have three or four membranes, they probably originated as photosynthesizing cells that were engulfed by another eukaryote before they were reduced to a plastid, with the extra membranes coming from the new host. This secondary endosymbiosis has happened at least three times: twice in green plants and once or more among red algae.

Mitochondrial Eve

Most eukaryotic cells contain two genomes, one in the nucleus and another in mitochondria (photosynthetic cells also have plastid DNA). In species that reproduce through sex, sperm carry DNA from the male nucleus, but eggs contain DNA from both female nucleus and mitochondria. In humans, that means mitochondrial DNA (mtDNA) is passed from mother to daughter on an unbroken maternal line. In 1987, geneticist Allan Wilson published a landmark study that compared mtDNA of 145 people from five geographic populations, showing they were all descended from a single individual that probably lived in Africa about 200,000 years ago. The media named her 'Mitochondrial Eve' (Wilson preferred 'lucky mother'). Mitochondrial Eve is the most recent common ancestor of everyone alive today but, unlike Eve from the Bible, was not the first woman – other women were alive at the time but did not leave any living descendants. Nowadays, mtDNA can be used in genetic tests to reconstruct someone's ancestry. The mitochondria that power sperm motility can enter an egg during fertilization, but they are usually broken down, so paternal mtDNA inheritance is rare.

The condensed idea
Complex cells incorporate the descendants of symbiotic bacteria

18 Respiration

Cells 'breathe' to generate energy, powering all the metabolic reactions that sustain life. In aerobic respiration, oxygen is used to burn carbohydrates, a process that makes so much energy that biochemists could not explain how it works – until Peter Mitchell proposed a radical new idea.

Anerobic respiration is sometimes described as a physiological process, the transport of oxygen (from air or water) to cells, with carbon dioxide going the other way, combining gas exchange with circulation. But this is a weird way to look at breathing as it ignores the *purpose* of respiration, like saying that the point of a battery is to keep it charged. Organisms breathe to generate energy. Breathing is a cellular process and how it works has been known since Peter Mitchell proposed his 'chemiosmotic' theory more than 50 years ago. Like many revolutionary ideas, chemiosmosis was initially opposed by the scientific community. It probably did not help that the eccentric Mitchell soon quit formal academia to work in a biochemical laboratory he had built while renovating a house in Cornwall, southwest England. Nevertheless, his single-minded vision was ultimately recognized in 1978, when he won a Nobel prize.

Generating energy

Aerobic respiration uses oxygen to burn food and produce the energy-carrying molecule ATP. Based on the reactions alone, this process, known as oxidative phosphorylation, makes an unexpectedly large amount of ATP. In the 1940s, biochemists believed equations had to be balanced (a technique called stoichiometry). When the ratio of ATP produced per oxygen molecule was measured, however, it turned out to be something like 2.5 – incompatible with the whole-number quantities expected.

Adenosine triphosphate (ATP) was isolated from muscle and liver extracts by German chemist Karl Lohmann in 1929, its structure – a DNA building block plus three phosphate groups – was proven by British chemist Alexander Todd in 1948. Between those two dates, Fritz Lipmann proposed that ATP is a universal energy carrier thanks to 'energy-rich phosphate bonds'. Enzymes that catalyse reactions often work as coin-operated proteins: ATP is inserted into a slot, then released as adenosine diphosphate (ADP), which sometimes leaves phosphate (P) behind. This pays the protein for performing a metabolic task like transporting a molecule across a membrane. ATP is used in so many biochemical transactions that it is known as the 'energy currency' of cells.

Food is based on carbohydrates – molecules containing carbon, hydrogen and oxygen – and digestion breaks down complex sugars, fats and proteins into simpler molecules for three biochemical pathways that generate energy-carrying molecules like ATP. The first pathway, glycolysis, occurs in the cell cytoplasm and does not require oxygen. It begins with glucose (a six-carbon sugar) and ends with two pyruvate (three-carbon) molecules. In complex cells, oxidative phosphorylation also takes place inside mitochondria, using a circular pathway known as the Krebs cycle (after Hans Krebs, who worked out its reactions in 1937).

> Not only can metabolism be the cause of transport, but also transport can be the cause of metabolism.
>
> Peter Mitchell

Most ATP, however, is produced by the third biochemical pathway: an electron transport chain on the inner membrane of mitochondria (or on the cell surface of bacteria). The first two pathways have exact reactions and stoichiometry; back in the 1940s there was no reason to suspect the electron transport chain would be any different. But biochemists were deceived: the chain typically produced fewer ATP molecules than predicted. While most tried to find intermediate reactions to make extra ATP, Peter Mitchell realized that aerobic respiration was not simply chemistry, but biology.

Proton gradients

An electron transport chain involves electrons passing from one protein molecule to another along a membrane. This occurs several times before a final 'electron acceptor' – oxygen – terminates the chain. This is why aerobic respiration requires oxygen. Mitchell proposed that transporting molecules across the membrane could also be coupled to the metabolic reactions that make ATP, and suggested a process similar to osmosis – the net flow of a chemical

During aerobic respiration, glycolysis splits food into pyruvate molecules that enter the Krebs cycle to ultimately produce large amounts of ATP. But if oxygen is not present, glycolysis must be a cell's main source of ATP. This occurs in overworked muscles, and in anaerobic organisms. The pyruvate made through glycolysis becomes a waste product, which is converted to other molecules, such as lactate in muscles, or ethanol and carbon dioxide in yeast. Man has been harnessing the latter anaerobic process – fermentation – for brewing and baking bread since the Stone Age. In the mid-19th century, scientists showed that fermentation is performed by cells. French microbiologist Louis Pasteur showed that bacteria turn milk sour, for example, which led to pasteurization. The German chemist Eduard Buchner performed alcoholic fermentation using extracts from yeast cells in 1896, the first time a complex biochemical process had been demonstrated outside living cells. Researchers such as Gustav Embden and Otto Meyerhof then studied the individual reactions and enzymes at each step so that, by 1940, the complete pathway of glycolysis was known. Along the way, a member of Meyerhof's lab, Karl Lohmann isolated the energy-carrying molecule ATP.

down a concentration gradient towards the more dilute side of a membrane. In his 'chemiosmotic coupling' theory, that chemical is a hydrogen ion (H^+).

Chemiosmosis therefore works like a hydroelectric dam, where higher pressure on one side forces water through turbines that capture energy from movement. In Mitchell's theory, proteins in the membrane – the 'dam' – are powered by the electron transport chain, enabling them to pump protons towards the outside of the mitochondrion and create a 'proton motive force', that effectively turns the membrane into a battery. In 1965, Mitchell and Jennifer

Moyle, the only colleague in his rural lab, backed up the theory by measuring the difference in pH (H^+ concentration) between the inside and outside of mitochondria. Meanwhile, other researchers had discovered the system's 'turbine' – an enzyme called ATP synthase.

In 1964, American biochemist Paul Boyer had proposed that this enzyme functions by changing shape, and in 1994 British scientist John Walker confirmed the structure involved. Around the same time, other researchers showed that one half of ATP synthase resembles a water wheel: the flow of protons through the wheel creates movement. As this molecular motor rotates and changes the shape of the enzyme, binding sites are repeatedly exposed to catalytic reactions that generate ATP. Aerobic respiration turns the membrane into a lucky slot machine: insert a few coins (in the form of electrons to pump protons across) and you win a jackpot of ATP every time.

The condensed idea
Energy is generated by the exchange of gases

19 Photosynthesis

The vast majority of life on Earth is ultimately powered by the Sun, thanks to organisms that make carbohydrates using carbon dioxide and sunlight. Their photosynthetic cells also release oxygen, which is vital for burning the energy-rich carbs that fuel metabolism.

Three billion years ago was a good time to be a chemoautotroph. Earth's atmosphere was thick with greenhouse gases like carbon dioxide, and chemoautotrophic organisms made food by capturing the energy released in chemical reactions. Then along came the first photoautotrophs. Floating at the ocean surface, bathed in sunlight, these solar-powered pioneers made carbohydrates after sucking CO_2 from the air. But they also released oxygen. What happened next was either one of the great events in life's history or a natural disaster worse than any mass extinction. The Great Oxygenation Event – or Oxygen Catastrophe, depending on your point of view – occurred about 2.3 billion years ago and ruined everything for chemically powered microbes. Oxygen is a sociable element that readily forms bonds with other molecules, making it toxic to chemoautotrophs. Poisoning their competition, photoautotrophs gradually altered the composition of air, which is now 21 per cent oxygen. Photosynthesis literally changed the world.

The carbon cycle

Oxygen is now vital to most life on Earth. Its importance was demonstrated by its discoverer, English clergyman and chemist Joseph Priestly, who in 1771 showed that a sprig of mint in an inverted jar would 'restore the air which had been injured by the burning of candles'. In 1779, Dutch physician Jan Ingenhousz showed that green leaves and stems only produce oxygen in light, and in 1782 Swiss pastor and botanist Jean Senebier suggested that plants absorb carbon dioxide and water to produce organic matter. These 18th-century observations give the basic formula for photosynthesis:

$$light + 2H_2O + CO_2 -> 2O + CH_2O + H_2O$$

As Melvin Calvin once said, 'If you know how to make chemical or electrical energy out of solar energy the way plants do it ... that is certainly a trick.' Traditional blue-black solar panels are made with silicon, but chemists are being inspired by nature to develop new sources of green energy. In 1988, Michael Grätzel and Brian O'Regan invented a photovoltaic cell from a thin film of titanium dioxide nanoparticles, made sensitive to light by dipping it in a coloured dye. This 'dye-sensitized solar cell' mimics how plants absorb photons with chlorophyll, uses low-cost materials and works in cloudy conditions to generate an electric current. Chemists are also copying the ability to produce fuel from light energy. Daniel Nocera has been working on an 'artificial leaf' that splits water into hydrogen and oxygen, while Nate Lewis of the US Joint Center for Artificial Photosynthesis is aiming to manufacture carbon-based fuels like methanol by fixing CO_2 in organic compounds, as in the Calvin cycle. These solar fuels could be transported to vehicles and other objects, replacing fossil fuels as a source of power.

The CH_2O in this equation represents carbohydrates – energy-rich molecules such as sugars, which most organisms use to power metabolism. Photoautotrophs supply carbs to heterotrophs, organisms that cannot make their own food, and both release CO_2 during respiration. Together with environmental processes such as gas exchange at the ocean surface, photosynthesis drives Earth's carbon cycle – the perpetual building and breaking of organic compounds.

Converting sunlight

Photosynthesis begins with energy conversion, a process performed by photosystems – pigments, proteins and other molecules that capture and convert energy from photons. The core component is the green pigment chlorophyll, isolated by French chemists Joseph Bienaimé Caventou and Pierre-Joseph Pelletier in 1817. When hit by photons, electrons in chlorophyll absorb energy, allowing them to break free of the molecule. This triggers a chain reaction where the energized electrons are passed between a series of molecules in an 'electron transport chain'. The chain creates NADPH and ATP, two molecules that will later release energy stored in their chemical bonds to power the synthesis of carbohydrates.

So that photosystems can repeatedly convert light to chemical energy, the electrons in chlorophyll must be replenished. This is achieved by photolysis (splitting molecules with sunlight), a reaction catalysed by the enigmatic 'oxygen-evolving complex'. For plants, the source of electrons is H_2O, and photolysis produces more O_2 than is needed for respiration – the excess is released as waste.

> **The essential features of the cycle with which we finally emerged were demonstrated on a wide variety of photosynthetic organisms, ranging from bacteria to the higher plants.**
>
> Melvin Calvin

The light-dependent reactions of photosynthesis actually involve two connected photosystems, as hinted by a single-celled alga, *Chlorella*. In 1943, American plant biologist Robert Emerson found that while the photosynthetic cells absorb light over a range of wavelengths, there is a drop in efficiency at the red end (680nm) of the spectrum. In 1957, he noticed that the rate of photosynthesis at red (680nm) and far-red (700nm) is enhanced when organisms are exposed to both wavelengths. This suggests two photosystems: electrons lose energy towards the end of one transport chain, but are re-energized at the start of the second photosystem.

Producing carbohydrates

Photosystems are embedded within membranes called thylakoids, which are folded to maximize the surface area exposed to light. In bacteria, thylakoids are an extension of the outer membrane, whereas plant and algal cells contain dozens of chloroplasts – capsule-shaped organelles dedicated to photosynthesis.

The NADP and ATP molecules, produced during the light-dependent steps of the process, are broken down to make carbohydrates by releasing energy from chemical bonds. In order to make food day or night, photosynthetic cells need a constant supply of carbon. Where does it come from? In 1945, scientists from the University of California at Berkeley led by Melvin Calvin used the radioactive isotope carbon-14 to track the path of carbon during photosynthesis in *Chlorella*, discovering a circular pathway of biochemical reactions that continually regenerates the same carbon-based compounds, a process commonly called the Calvin Cycle.

The cycle starts with CO_2 from air being attached to a simple sugar with five carbon atoms, RuBP, which produces a six-carbon molecule that splits into two 3PG, a three-carbon sugar. Plant cells transport one molecule of 3PG from the chloroplast to the cytoplasm, where it can be used to make complex carbs such as glucose. The other 3PG molecule goes through several steps where enzymes prompt NADPH and ATP to donate hydrogen and phosphate atoms, which eventually regenerates RuBP to restart the cycle. During the first step, CO_2 – a volatile gas – is fixed to a stable molecule, or 'carbon fixation'. This step is catalysed by RuBisCO, an enzyme that constitutes 30–50 per cent of the soluble protein in leaves, and is probably the most abundant protein on the planet.

The condensed idea
Solar energy is captured to make food

20 Cell division

All cells arise through division. While simple organisms like bacteria can split by binary fission, division in complex cells is complicated by the nucleus and numerous chromosomes, which means cells divide through the multistage process of mitosis.

The history of cell theory – the idea that all life is made of cells – is one of 19th-century scientists stealing or ignoring each others' ideas. While Germans Matthias Schleiden and Theodor Schwann are often credited as its 'discoverers', many could claim the title. Polish researcher Robert Remak, for instance, observed division in animal cells and disproved Schleiden and Schwann's belief that they arise spontaneously from crystals.

Cell theory owes a debt to drawing and colouring. The study of division under a microscope was made easier with dyes such as indigo, which stained structures inside the nucleus, as discovered in salamanders by German biologist Walther Flemming. In 1878, he named the structures 'chromatin' – today, we call them 'chromosomes' ('coloured bodies' in Latin). He noted that chromosomes are either a cloudy mess or appear as threads, or 'mitosen', the states that now define two periods in a cell division cycle: interphase, when a mother cell is growing; and mitosis, when its chromosomes are split between two daughter cells. Biologists identify five distinct phases during mitosis.

Mitosis

After DNA has been duplicated during interphase, the first stage of mitosis – prophase – begins. Clouds of genetic material condense to form distinct chromosomes called 'sister chromatids' that resemble pairs of stripy socks and are joined in the middle to form X-shaped structures. The condensation reaction is catalyzed by DNA-binding proteins, such as the appropriately named condensin, which forms coils around chromosomes so they become thousands of times more compact.

Before the second stage – prometaphase – a structure called the centrosome divides and migrates to opposite poles of the cell, and a scaffold of microtubules – the spindle – grows between its parts. As seen in 1848 by botanist Wilhelm Hofmeister, the nuclear envelope

Mitosis (left) involves one round of cell division, while meiosis (right) has two. During mitosis, all chromosomes in the parent cell align in single file, as sister chromatids, before being split for daughter cells with pairs of chromosomes (diploid). In meiosis I, homologous pairs from each parent align side-by-side and exchange genetic material through recombination before separation. In meiosis II, the chromatids are separated to leave one set (haploid) in egg or sperm cells.

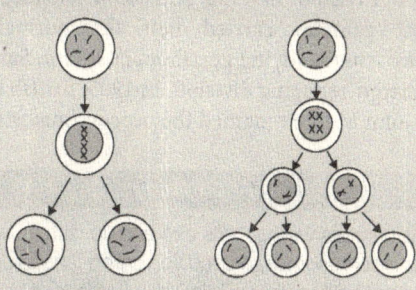

breaks into bubble-like vesicles or 'klumpen', allowing chromosomes to become attached to the microtubules. In 1888, Theodor Boveri saw that chromosomes align in single file along the spindle. At this halfway point – metaphase – there is a brief pause, as if the cell is taking a deep breath before taking the plunge of dividing.

During anaphase, sister chromatids in each chromosome are pulled towards opposite poles by a tug-of-war across the spindle. The molecule that keeps those chromatids together, cohesin, is now digested by enzymes. At the final stage of mitosis – telophase – sister chromatids reach the poles, the spindle falls apart and vesicles reassemble into nuclear envelopes around two sets of chromatids, which de-condense, unravelling the stripy socks of chromosomes. The cell then splits in two by a final process called cytokinesis.

Meiosis

Sexual reproduction usually combines a set of chromosomes from each parent. Belgian zoologist Edouard van Beneden was first to realize what this meant for cell division. Studying fertilized eggs of the nematode worm *Ascaris megalocephala* in 1883, Van Beneden said: 'Each pronucleus is equivalent to a half-nucleus endowed.' We now say gametes

> No mode of cell multiplication other than that involving cell division with indirect nuclear reproduction has been demonstrated up to now.
> Walther Flemming

– sperm or egg – with one set of chromosomes are 'haploid' and cells with two sets are 'diploid'. Producing gametes therefore requires a special kind of division. In 1902, American scientist Walter Sutton realized that gametes carried half the normal number of chromosomes. After studying grasshopper sperm, Sutton concluded that they undergo 'reducing division'. In 1905, British biologists John Farmer and John Moore renamed the process 'maiosis' (meiosis).

Aneuploidy

When division goes wrong, cells can end up with an abnormal number of chromosomes, a condition called 'aneuploidy'. In many species, body cells are 'diploid' with pairs of chromosomes (one set inherited from each parent), but errors in cell division can lead to the loss or gain of chromosomes. One way this can happen is because a chromosome does not attach to the spindle properly. It does not get dragged to opposite poles of the cell and then effectively 'disappears' by being outside the nucleus, where its genes are not read. Alternatively, a pair of sister chromatids might not separate during meiosis, a phenomenon known as 'non-disjunction'. This can leave a gamete (sperm or egg cell) with no copies of a certain chromosome so that, after fertilization, the embryo has only a single copy from one parent, a condition called monosomy. A gamete can also carry two copies, meaning the resulting embryo inherits three copies, or a 'trisomy'. In humans, Downs syndrome is caused by trisomy of chromosome 21, while women with Turner syndrome have a single X chromosome. In aneuploidy, an imbalance in the number of genes produces the wrong 'dosage' of proteins, disrupting a cell's delicate biochemistry.

Meiosis differs from mitosis in two main ways: it involves two rounds of division, and chromosomes are separated differently – mitosis produces two diploid cells, while meiosis makes four haploid gametes. During the first round – meiosis I – pairs of maternal and paternal chromosomes align side-by-side and then separate, rather than (in mitosis) single-file with splitting of sister chromatids. This splitting occurs in meiosis II. Based on the side-by-side pairing of meiosis I, Sutton suggested that chromosomes carry genes. He also pointed out that cells do not have a maternal side and paternal side – each chromosome pairs at random, so they can be inherited in an array of combinations. For three pairs, there are 2^3 (8) possible arrangements at metaphase. For humans, excluding the sex chromosomes, there are 2^{22} or over four million combinations.

Binary fission

Without a complex nucleus or numerous chromosomes to worry about, cell division is much simpler and faster. A cell grows to twice its size then splits in two. In rod-shaped bacteria and most other prokaryotic organisms, once the cell reaches twice its length, it divides by binary fission. Bacteria have a single circular chromosome attached to the outer membrane about halfway along the cell. Replication starts from an 'origin' on each strand of the double helix in both directions, producing two DNA loops. In 1991, microbiologists Erfei Bi and Joseph Lutkenhaus showed that a molecule called FtsZ creates a 'Z-ring' structure that contracts like the drawstring of a bag or purse – splitting the cell in a similar way to cytokinesis in eukaryotes.

The condensed idea
Chromosome duplication complicates cell division

21 The cell cycle

Division is the most significant event in the life of a cell. Like parents preparing for the birth of a baby, a mother cell wants events to run smoothly when it splits into two daughters. In the complex cells of eukaryotes, everything from budding yeast to the blue whale, this is achieved through the cell division cycle.

Cell size is limited, so a bigger body means more cells. The largest animal that has ever lived, the blue whale, has almost 100 quadrillion cells, all of which come from a single fertilized egg. An adult human male has 36 trillion cells and billions are replaced every day. Cell division leaves a lot of room for error – lose control and you might end up with cancer. To minimize the possibility of things going wrong, eukaryotic cells check the conditions at several stages before dividing.

Phases

Eukaryotic cells are 'born' via division and 'die' when they divide. This life cycle consists of four phases. Each cell grows larger during the first gap phase (G_1), DNA replication occurs during synthesis (S), a second gap phase (G_2) checks that genetic material has been copied, while mitosis (M) distributes the duplicated chromosomes between two nuclei. One mother cell then splits into two daughter cells. A cycle can take minutes or days, depending on the cell – in humans it lasts one day on average, with the first three phases, known collectively as 'interphase' – taking up most of the time. Some cell types, including neurones and heart muscle, never complete a cycle and instead enter a resting state: G_0. Cells cannot go backwards through the cycle, as shown by cancer researchers Potu Rao and Robert Johnson in 1970.

Checkpoints

Cancer cells behave unusually, however. In 1965, microbiologist Donald Williamson used radioactive labelling to track DNA synthesis during growth of baker's yeast (a unicellular organism) and showed that its phases matched those in a multicellular body, meaning it could be used as a model for studying the eukaryotic cell cycle in general. Mutations that block cell division would normally end a life cycle, but Hartwell's mutants were sensitive to temperature, growing normally at 25°C

(75°F), but not at 36°C (96.8°F). This provided an on/off switch for division. Different strains stopped at different points, so each mutation – and its gene – could be assigned to a phase. Hartwell described dozens in 1970, but the most interesting gene was 'cell division cycle 28' or CDC28 – nicknamed 'start' because it determines whether a cell enters the G_1 phase. This led to the idea of checkpoints.

After reading Hartwell's research, British geneticist Paul Nurse also got interested in yeast so went to Edinburgh University to learn from zoologist Murdoch Mitchison, who studied the fission yeast *Schizosaccharomyces pombe*. Nurse also found temperature-sensitive strains, including mutants that would rush through a cell cycle. These yeasts divided prematurely and were below normal size so, in a nod to the Scottish word for 'small', they were named 'wee' mutants. In 1975, Nurse showed the CDC2 (wee2) gene controls whether a cell passes the G_2 to M checkpoint.

CDC2 is central to the cell cycle. In 1982, Nurse used a technique called 'cross-species complementation' to identify genes that rescue the ability to pass checkpoints. This involved inserting different CDC genes into modified DNA and giving them to temperature-

Phases of the cycle

The cell division cycle has four phases: G1 involves growth, S is for DNA synthesis, the G2 phase checks genetic material was copied, M distributes duplicated chromosomes between two nuclei. After M phase (mitosis), a mother cell splits into two daughter cells. Some cells exit the cycle and enter a resting state: G0.

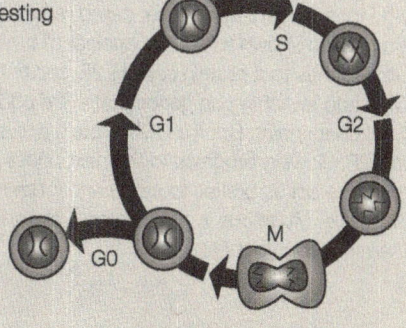

sensitive mutants to test whether they would then grow. Cross-referencing the species and genes, Nurse discovered that the CDC2 gene in fission yeast has a similar function to CDC28 in budding yeast, the gene controlling 'start' in a cell cycle. This is impressive given that while both are unicellular organisms, their common ancestor lived over a billion years ago. In 1987, Nurse, then working at a cancer research institute in London, used the same technique to show that human DNA could also rescue mutant yeast, thereby identifying CDC2 in humans.

Cycles

The CDC genes make proteins that control progress through a checkpoint. If those proteins were always active, cells could bash through the barriers. What controls the controllers? While studying eggs from the sea urchin *Arbacia punctulata* in 1982, British

Cancer control

Abnormal cells break through barriers that block growth and division, including cell cycle checkpoints, so cancer-stopping proteins called 'tumour suppressors' will check whether a cell should progress from one phase to the next. One key suppressor is p53 (tumour protein 53), which scans DNA for damage – hence its nickname: 'guardian of the genome'. If it detects damage, p53 changes shape and turns on various genes, including ones which make proteins that repair the genetic material. Another key suppressor is the Rb protein. Its gene was identified in children who inherit a rare eye cancer called retinoblastoma, but it has since been shown to protect against all tumours. When active, Rb attaches to a protein called E2F, preventing E2F from binding DNA and switching on genes that – like p53 – lets cells pass the G_1/S checkpoint. But if a CDK protein activates Rb, it cannot stick to E2F, allowing progress to the next phase. Tumour suppressors therefore act as brakes to stop cancer driving through the cell cycle. Mutations in p53, Rb or cyclin and CDK genes can therefore contribute to cancer.

biochemist Tim Hunt showed that some proteins are made and degraded during every cell cycle. As their levels rise and fall periodically, he named them 'cyclins', and suggested they might be related to MPF or 'maturation-promoting factor' – a molecule that scientists had spent two decades trying to find.

In 1988, several researchers, including Paul Nurse, discovered that the elusive MPF is a combination of two proteins: cyclin B and CDC2. Nowadays, a CDC protein is known as a CDK or 'cyclin-dependent kinase' (CDC2 is CDK1 in humans). Kinases are enzymes that activate other proteins by adding phosphate to them, which explains how a cyclin-CDK pairing controls metabolism: the CDK alters other proteins while levels of its partner cycle up-and-down, ensuring the CDK is not always active. For example, cyclin E levels rise during the G_1 phase to prompt DNA synthesis, then drop during S phase.

> Perhaps, not unreasonably, some wondered what a yeast researcher was doing in a cancer research institute.
> Paul Nurse

Tumour suppressor proteins such as pRB and p53 are active by default to help prevent cancer, but are deactivated by specific cyclin-CDK pairs when conditions look clear for cell division. Yeasts have one cyclin and one CDK whereas humans have about a dozen of each. However, the fact that these species are separated by billions of years of evolution shows that the control system originated early in the history of eukaryotic organisms, reinforcing the fact that, whether budding yeast or blue whale, the cycle is vital to the division of complex cells.

The condensed idea
Complex cells regularly check that division is running smoothly

22 Cancer

All animals probably suffer from out-of-control cells, which can lead to cancer. In humans, according to the World Health Organization there were an estimated 19.3 million cases (and 10 million deaths) in 2020 – an annual figure that is predicted to rise to 30.2 million cases in 2040.

Cancer occurs when a body loses control of its own cells. It can originate from any tissue as a group of abnormal cells – a tumour – that spreads to make over 100 different diseases with similar features. Hallmarks of cancer include: uncontrolled growth and division, gaining independence and immortality and manipulating tissues to migrate around the body.

Tumours are caused by mutations – either permanent DNA changes (genetic mutations) or reversible modifications to the genome (epigenetic mutations). Risk increases with age because mutations stack up over time. Some arise spontaneously, others are inherited, but many are triggered by environmental factors known as carcinogens. Ultraviolet light, which triggers skin melanoma, is a physical carcinogen, while tobacco is a chemical carcinogen. A biological carcinogen was discovered in 1910 by American virologist Peyton Rous, who took a tumour from a Plymouth Rock hen and put it in other chickens. The cancer continued to grow, even when Rous filtered the tumour to remove cells.

Growth and division

Rous sarcoma virus revealed why cancer cells have uncontrolled growth. In 1976, American biologists Michael Bishop and Harold Varmus detected the viral 'Src' (sarcoma) gene in normal chicken cells, suggesting that the virus had stolen a cellular version of Src, which mutated into a viral form that triggers tumours. Two years later, Bishop and Varmus found Src in humans and mice, and also showed that the viral gene encodes a 'kinase' – a protein that activates others. In 1979, they isolated the Src protein from uninfected chicken, quail, rat and human cells.

Src is a 'proto-oncogene' – in other words, a gene that contributes to cancer when mutated, becoming an 'oncogene'. These genes often encode proteins from a signalling pathway, a domino-like chain of molecules that relays instructions from the body. Messages are received by a cell's surface receptors, which transmit the signal to the nucleus, where gene activity alters behaviour. In the 1980s, researchers found that viruses also steal genes for growth factors and matching receptors, further highlighting the importance of signalling. In glioblastoma, for example, brain cells stimulate themselves by releasing platelet-derived growth factors. Without growth signals, cancer becomes self-sufficient.

While molecular biologists focused on infections, geneticists looked at patterns of inheritance. Studying populations in 1954, Peter Armitage and Richard Doll proposed that cancer requires at least two mutations. In 1971, Alfred Knudson examined the family history for 48 cases of an eye tumour, retinoblastoma, and his statistics supported what became the 'two-hit hypothesis'. This is explained by the fact we inherit a copy of a gene from each parent: children with retinoblastoma get two faulty copies, but cancer occurs if your only working copy mutates later in life. The retinoblastoma gene (Rb) is a 'tumour suppressor' that stops uncontrolled cell division. The most crucial suppressor is p53 (tumour protein 53), discovered by Arnold Levine and David Lane in 1979. Nicknamed 'the guardian of the genome', p53 alerts another protein, p21, to halt the cell-division cycle if damage is detected. Half of all tumours have a mutated p53 gene, enabling cells to continuously divide. Cancer is therefore insensitive to anti-growth signals. Mutated oncogenes are like a stuck accelerator, mutated tumour suppressors are a broken brake pedal and handbrake – together they keep the defective car of cancer hurtling forwards.

Independence and immortality

When something goes wrong in normal cells, they commit suicide or 'apoptosis'. But in many tumours, mutated p53 fails to block the activity of another protein, Bcl-2, that stops mitochondria telling enzymes to destroy the cell. In 1988, biologist David Vaux found that adding Bcl-2 genes to blood cells helped them survive without growth factors, showing that survival and growth are controlled separately. Tumours with mutated Bcl-2 evade suicide.

Cancer cells can also live forever. Normal cells have a finite number of divisions, dictated by 'telomeres' – dispensable DNA at the end of chromosomes that get shorter each time a cell divides, showing its age and prompting it to stop replicating. Cancer gets round this with cosmetic surgery: about 90 per cent activate genes for the enzyme telomerase, which adds telomeres to chromosomes, tricking the cell into thinking its DNA looks younger than it is.

Manipulation and migration

Plants do not get cancer. They can develop abnormal outgrowths, but because their cells are restricted by rigid walls, a tumour has limited potential to invade surrounding tissues. In animals, cancers can manipulate cell flexibility. American surgeon Judah Folkman suggested that cancer stimulates the formation of new blood vessels – angiogenesis – a process that normally occurs when the body builds tissue or heals wounds. Manipulating the oxygen and nutrient supply feeds ever-hungry cancer cells.

A tumour only becomes cancer after migrating from its point of origin, going from benign to malignant. The process, 'metastasis' ('displacement' in Greek), begins with cells breaking through a basement membrane. The cells then squeeze through the wall of a capillary to enter the circulation, which can involve changing shape (anaplasia). Cells are transported via the bloodstream or lymphatic system and become stuck in a distant capillary. Those that survive the hostile new environment can invade and colonize surrounding tissue, growing into a metastatic tumour. Metastasis is behind 90 per cent of cancer-related deaths.

> Exploration of the cancer cell is akin to archaeology: we must infer the past from its remnants in the present, and the remnants are often cryptic.
>
> Michael Bishop

Development of cancer is an evolutionary process. Uncontrolled proliferation, which starts with a single abnormal cell, creates a population of clones. During clonal evolution, some cells will carry mutations that help 'individuals' resist environmental challenges – including attacks by the immune system, radiation or chemotherapy. Any surviving cells reproduce and, over the repeated rounds of mutation and natural selection, a tumour evolves the abilities that make cancer so hard to beat.

Contagious cancers

Devil facial tumour disease (DFTD) is a transmissible cancer that infects Tasmanian devils. In 2012, geneticist Elizabeth Murchison found that DFTD originated in a single female devil, then accumulated mutations so that parasitic cancer strains now have DNA distinct from each other and their host. The tumours form lesions and lumps that interfere with feeding, leading to starvation, and are transmitted by biting the face during fights. Dogs can catch canine transmissible venereal tumour (CTVT), a sexually transmitted disease that incubates for months before symptoms appear, enabling its spread. While CTVT has become less virulent over 11,000 years of evolution, DFTD is only about 20 years old and has devastated the devil population – an 80 per cent decline since the first case was identified in 1996, with the potential to drive the species to extinction by 2035. Devils, dogs and Syrian hamsters are the only animals known to catch cancer, suggesting it is very rare in nature.

The condensed idea
Body cells behave as selfish individual organisms

23 Viruses

They cause the deadliest diseases known to man, including AIDS, Covid and smallpox – but humans are just one of many species that suffer from viruses. These tiny parasites are undoubtedly the world's most successful life forms, and can infect every branch in the tree of life, from plants to bacteria.

To appreciate viruses, you temporarily forget the fact they are pathogens that sometimes cause disease and death to their hosts. They have climbed Mount Everest while being inside a human body, and are found on the sea floor. They are Earth's most abundant life form, with the global total number of viruses estimated at 10 nonillion (10 followed by 30 zeroes) – three times more than bacteria and archaea combined. We live on a planet of viruses.

Smaller than cells

Large viruses like *Variola* (smallpox) can be seen under a light microscope, but most need electron microscopes, which were developed by Ernst Ruska in the 1930s. In 1939, Ernst's brother, German physician Helmut Ruska, invented a 'shadowing' technique – bouncing electron beams off things coated in heavy metal atoms like uranium – to capture the first images of viruses, revealing their size. Imaging became much easier in 1959, when biologist Sydney Brenner and physicist Robert Horne developed a 'negative staining' technique using carbon and metal salts. We now know that viruses are on average 100 times smaller than a bacterium.

The tiny nature of viruses was first revealed in one species that almost single-handedly established the field of virology. In 1886, German scientist Adolf Mayer found a 'mosaic disease' that forms spots on tobacco leaves. It could be transmitted to healthy plants after using paper to filter a liquid extract and Mayer assumed it contained bacteria. Russian botanist Dmitri Ivanovsky used a porcelain filter with pores narrower than a micron (one-thousandth of a millimetre) to exclude cells from sap, but concluded the disease was caused by a toxin. In 1898, Dutch microbiologist Martinus Beijerinck described numerous tests on the mysterious disease, which included drying and storing filtered sap. Beijerinck suggested it was caused by an unknown virus, what he called a 'contagium vivum fluidum' - a contagious living fluid - that would only infect dividing cells. The infectious agent, Tobacco Mosaic Virus (TMV), was purified by Wendell Stanley in 1935. The American biochemist thought his protein crystals were contaminated with phosphorous but a year later, British virologists Norman Pirie and Frederick Bawden found the 'contamination' actually came from RNA.

Virion particles

The inside of a 'virion', an individual virus particle, was unveiled after the discovery that nucleic acid – DNA and RNA – is life's genetic material. In 1955, German biochemist Heinz Fraenkel-Conrat and American biophysicist Robley Williams showed that mixing viral RNA and bits of protein was enough to create Tobacco Mosaic Virus. That same year, British crystallographer Rosalind Franklin, whose X-ray crystal images helped reveal the double helix, found that TMV was rod-shaped. In 1956, she proved how the hollow rod contains RNA. Meanwhile, German biophysicists Alfred Gierer and Gerhard Schramm showed the RNA was infectious, suggesting it was TMV's genetic material. And in 1962, molecular biologists Marshall Nirenberg and Heinrich Matthaei showed that adding viral RNA to the contents of cells (in a test tube) would produce proteins, showing that the RNA has protein-coding genes.

> Only those organs of the plant that are growing and whose cells are dividing are capable of being infected; here only does the virus reproduce itself.
>
> Martinus Beijerinck

A virion has two or three parts: a genome, a coat and sometimes a wrapper. These all originate from molecules stolen from a host cell. Genetic information is carried by nucleic acid (either DNA or RNA), while the coat or 'capsid' is made from protein and may be an icosahedral structure, as used by rhinoviruses that cause the common cold, or a rod-shaped helix, as in TMV. Severe Acute Respiratory Syndrome Coronavirus 2 (SARS-CoV-2, which caused the Covid-19 pandemic) and Human Immunodeficiency Virus (HIV) have wrappers made from the membrane or nucleus of host cells, a lipid envelope embedded with proteins that enable invasion.

Infection

Viruses are intracellular parasites, exploiting their host's molecular machinery to replicate. When a virus reaches a cell, proteins on its capsid or envelope attach to receptors on the cell's membrane, unlocking its doors. Infective ability depends on matching molecules, so HIV can only invade white blood cells carrying the CD4 receptor. With viruses like HIV, the viral envelope and cell membrane fuse, allowing a capsid to penetrate. If cells have walls, a capsid might enter via pores or holes. Virions might include enzymes that degrade the capsid, a process called 'uncoating'. The naked viral genome then transforms the cell into a virus-making factory.

Replication varies according to the viral genome, which may be DNA or RNA, single or double stranded, positive or negative. In every combination, genetic material is copied and proteins are made using a

Virion structures

The genome of a virion (individual virus) is contained within a protein-based coat or 'capsid', which has two main forms. Helix structures are rod-shaped, as in Tobacco Mosaic Virus, or form a flexible filament, as in ebola. Icosahedral structures, including coronaviruses, resemble a football, while bacteriophages combine an icosahedral head, helical tail and 'legs'. The T4 bacteriophage's syringe-like tail injects the genome the cell wall of its host, *E. coli*.

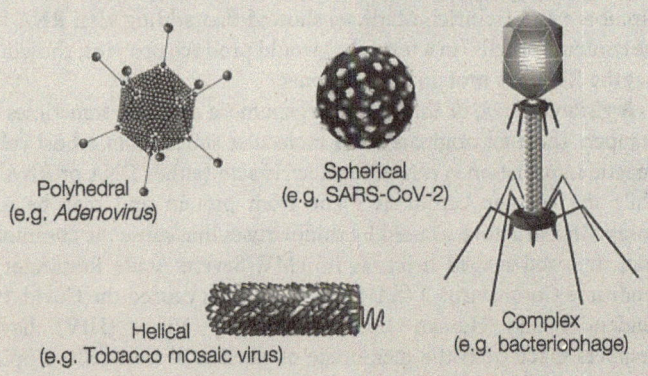

Polyhedral
(e.g. *Adenovirus*)

Spherical
(e.g. SARS-CoV-2)

Helical
(e.g. Tobacco mosaic virus)

Complex
(e.g. bacteriophage)

cell's gene expression machinery, but some viruses also bring special enzymes. In 1970, American geneticists Howard Temin and David Baltimore independently discovered that RNA tumour virus virions carry an enzyme, 'reverse transcriptase', that reads single-stranded RNA and copies it back to double-stranded DNA – a process that does not happen in normal cells. HIV produces reverse transcriptase, as well as another enzyme, 'integrase', which inserts a DNA copy of its RNA into the human genome, creating a 'provirus' that can remain dormant for years before causing Acquired Immune Deficiency Syndrome (AIDS).

Disease

Some viruses trigger mutations when viral DNA is inserted into the genome, as in cancer-causing oncoviruses, while others do little direct damage, sleeping while the cell divides or budding from its outer membrane, as in HIV. But many viruses replicate until their virions cause a cell to burst or 'lyse' – not deadly for a multicellular organism, but fatal for unicellular microbes. Parasites only cause severe problems if they reduce their host's ability to survive or reproduce. The common cold – a syndrome of viral infections in the upper respiratory tract, caused by over 200 different viruses – is not fatal, but by the time you recover, you have already helped spread the cold. Viruses like SARS-CoV-2 can even spread without symptoms.

The condensed idea
Parasitic life forms replicate by exploiting host cells

24 Prions

Prions cause contagious brain diseases in mammals. Unlike infectious agents such as viruses, they are not associated with nucleic acid; they are just proteins. They are not alive, so cannot be killed – they are also difficult to destroy, and there is no known cure. And yet, prions might not be all bad.

The highlands of Papua New Guinea, 1957: a mysterious disease is spreading through the Fore tribe. Victims have trouble standing, and suffer from spontaneous laughter and uncontrollable tremors. The locals call it kuru, or 'shaking'. District medical officer Vincent Zigas and Harvard doctor Carleton Gajdusek fail to find the cause. Toxins in the diet or environment are ruled out, but kuru is more common within families (an inherited disorder?) and the shaking suggests brain disease. Despite a lack of inflammation, Gajdusek suspects viral encephalitis. He sends samples home.

Sponge brains

Under the microscope, slices of tissue from kuru victims are full of holes, giving the brain a spongelike appearance. After seeing images at an exhibition in 1959, William Hadlow, a veterinarian specializing in pathology, noted similarities to scrapie, a fatal disease in sheep that causes symptoms including itching, lost coordination and paralysis. Scrapie was discovered in the 1700s and studied in the 20th century, when scientists demonstrated it could be transmitted to goats and mice.

Scrapie and kuru are transmissible spongiform encephalopathies. Another example, BSE (bovine spongiform encephalopathy) or 'mad cow disease', was identified in 1986 as an epidemic in British cattle, originating from meat-and-bone meal contaminated with scrapie. In humans, Creutzfeldt-Jakob Disease (CJD) arises as sporadic CJD, through medical contamination, or familial CJD from inherited mutations. A fourth type appeared in 1996: vCJD (variant CJD), caught by eating BSE-infected beef. The mystery of kuru

> I had anticipated ... a small virus and was puzzled when the data kept telling me that our preparations contained protein but not nucleic acid.
> Stanley Prusiner

had actually been solved decades before, mirroring BSE and vCJD: the Fore tribe engaged in ritualistic cannibalism that honoured dead relatives by cooking and eating them. Once the tradition stopped, so did the spread of kuru.

Slow virus

In the 1960s, experiments by Carleton Gajdusek and Joe Gibbs showed a long incubation period between kuru infection and symptoms, which they explained by an enigmatic 'slow virus'. Scrapie researchers also believed in a viral pathogen – a virus so difficult to destroy that it still caused disease after being stored in formaldehyde. The infective agent was also resistant to the usually effective treatments of heat and UV radiation.

Then in 1972, Stanley Prusiner began a residency in the neurology department at the University of California San Francisco. Here, he studied CJD using hamsters, and isolated the 'slow virus' behind it. But he repeatedly failed to detect any genetic material – only protein. In 1982, Prusiner made the controversial suggestion that scrapie is caused by 'small proteinaceous infectious particles that are resistant to most procedures that modify nucleic acids'. 'Proteinaceous infectious particle' was soon shortened to 'prion'.

Infectious proteins

How could proteins be contagious? Theoretical scientist John Griffith proposed three plausible mechanisms in 1967. One was self-replication, where an infective agent is an 'aberrant form of protein that spontaneously got made, and could serve as a template to induce production of more aberrant forms'. In 1985, Prusiner and Swiss biologist Charles Weissmann isolated the gene encoding the prion from a scrapie-infected hamster brain, and showed that the gene could be detected in normal hamsters, mice and humans. This proved that a prion's 'genetic' material occurs in cells. The abnormal prion protein is named PrP^{Sc} (Sc for scrapie) and the normal form is PrP^{C} (C for cellular). In 1992, Prusiner and American scientist Fred Cohen used computer tools to predict protein folding, which suggested that PrP^{Sc} secondary structures are mainly flat 'beta-sheets', while PrP^{C} consists of springlike 'alpha-helix' structures. A prion is a misfolded protein that prompts the normal form to change shape.

Like prion diseases, disorders like Alzheimer's and Parkinson's are associated with abnormal aggregates that accumulate with age and reduce cognitive ability. Unlike prions, proteins that form aggregates are not infectious agents – but they do have similar features. In Alzheimer's disease, the most common cause of human dementia, some regions of the cerebral cortex shrink in size as aggregates appear: the peptide (protein fragment) called 'beta-amyloid' creates amyloid plaques, while the 'tau' protein forms twisted, neurofibrillary tangles inside neurones, which block a transport network inside the cells. When beta-amyloid or tau are transferred from one neurone to another, they can 'seed' a new aggregate, which is why they are called 'prion-like' proteins – they seem to be misfolded versions that convert normal proteins into abnormal structures. 'Self-replication' might be causing an important supply of proteins to run out, aggregates themselves might be causing complications (as in prion disease) or both might be a problem. The cause of the initial misfolding is not known, and it is also not clear what triggers neurodegenerative diseases in the first place, as only 10 per cent are linked to genetic changes. As exposure to trauma and other environmental factors increases over time, it might simply be due to old age.

Abnormal agreggates

Prions are the vampires of molecular biology. Replication of nucleic acid involves using a single strand of DNA or RNA as a template. The prion, by contrast, does not create a copy of itself from scratch, but uses its own shape as a template to transform an existing PrP^C protein into the vampiric PrP^{Sc} molecule. Chain reactions then prompt more and more proteins to accumulate.

The beta-sheet of PrP^{Sc} is flatter than PrP^C, allowing prions to form stacks that become abnormal aggregates of 'amyloid' fibres. These toxic fibres kill neurones, prompting astrocyte cells to clear debris and leave holes in the brain – spongiform disease. Because prions are so tough, fibres remain while the astrocytes and holes increase.

One important piece of evidence to support the protein-only hypothesis came in 2004, when Prusiner's lab got bacteria to make prions that developed into amyloid fibres, which caused neurological dysfunction when injected into mice. Amyloid fibres can cluster to form deposits of amyloid plaques, and a similar process occurs in neurodegenerative diseases.

Memory molecules

But what does normal prion protein do? In 1992, Weissmann found that PrP^C is a receptor on cell membranes. Genetically modified mice without the gene were also immune to scrapie infection and stayed healthy for months, suggesting the protein is dispensable. But in 1999, two teams of Japanese researchers discovered that when mice lack PrP^C, their brain cells lose the myelin sheath that insulates electrical activity, and neurones called Purkinje cells die. It seems that normal prion proteins actually protect the brain.

Beneficial prions have since been discovered in many species. Dozens occur in budding yeast, for instance. In 2003, neurobiologists Kausik Si and Eric Kandel found the protein CPEB (cytoplasmic polyadenylation element binding) in a sea slug. When CPEB binds to mRNA transcribed from a neurone's genes, that cell produces proteins, which is required for memory storage. Remarkably, one end of CPEB resembles a yeast prion. When Si and Kandel put CPEB into yeast cells, proteins turned into prions. If a brain cell deliberately turned CPEB into a prion, it would keep producing proteins. Self-replicating proteins might therefore explain how long-term memory is maintained – or cause loss in diseases like Alzheimer's.

The condensed idea
Self-replicating proteins can damage or protect the brain

25 Multicellularity

The first major transitions in the history of complex life were the development of the nucleus and mitochondria in eukaryotic cells. These events probably happened only once, whereas the evolution of a multicellular body has occurred multiple times, suggesting it offers many advantages.

A solitary, single-celled organism performs every vital task – including locomotion, protection and reproduction – on its own. It is a jack-of-all-trades, master of none. A multicellular body, in contrast, can divide the labours of life between specialized tissues. The most basic division is two types of cell: reproductive cells for transmitting heritable information to the next generation, and body cells for everything else.

Division of labour

The distinction between reproductive 'germ' and 'soma' – body cells – was first proposed in 1883 by German zoologist August Weismann, who suggested that this division of labour enabled organisms to evolve complex bodies. Through differentiation, soma gives rise to cells for different jobs, specialized for everything from feeding to photosynthesis. An organism is multicellular if its cells are specialized, stick together, depend upon each other and communicate. Without these four features, a 'body' is just a colony of cells. The ancestors of living species are no longer around, however, so origins get a bit more leeway. Based on two characteristics – cell adhesion and communication – multicellularity may have arisen ten times: once in the animal kingdom, three times in fungi and six in plants.

Sticking together

How does multicellularity originate? The events that gave rise to modern bodies happened millions of years ago, so scientists often compare single-celled species to multicellular relatives. The perfect model system is the Volvocaceae, a family of green algae whose members go from single-celled organisms to having bodies containing thousands of cells. Scientists usually compare two species: *Chlamydomonas reinhardtii*, a unicellular alga that absorbs its

Biofilms

Bacteria often form a biofilm, an extracellular matrix that holds cells together in a kind of slime. This microbial mat is made from sugars, proteins, lipids and nucleic acids – material released by cells after they burst in a stressful environment. This prompts nearby organisms to change their genetic activity and therefore their features and behaviour. A biofilm creates a barrier that lets a collection of cells share metabolites but mostly blocks entry of toxic substances. Biofilms formed from *Staphylococcus aureus* and *E. coli* contain cells that are more resistant to antibiotics, for example, which has implications for stopping superbugs like MRSA. Films can potentially form on most surfaces, from a thin layer at the air-water interface to Petri dishes in the lab. Although a biofilm has many features of multicellularity (such as sticking together) and cells can share benefits like defence against predators, it is a transient form. Unlike a multicellular body, biofilms are not always made up of cells from the same origin. They might not even be from the same species. As a result, there is far more conflict of interest and competition within the 'community', leaving it vulnerable to the rise of cheating cells, ultimately making the mats unstable.

whiplike flagella before division, and *Volvox carteri*, which has about 16 large reproductive cells inside a transparent sphere – a gel-like matrix with 2,000 tiny body cells, each with flagella to steer the spherical body towards sunlight for photosynthesis.

Studying mutant strains of *Volvox*, developmental biologists have found several genes that control how the alga makes large germ and smaller somatic cells, or 'gonidia'. In 1999, Stephen Miller and David Kirk discovered glsA (gonidialess A), a gene required for asymmetric division – a mutated glsA makes equally sized cells. In 2003, Miller isolated the equivalent gene from the unicellular *Chlamydomonas* and gave it to mutant *Volvox*, causing the multicellular organism to regain different-sized cells. Geneticists led by Daniel Rokhsar compared the genomes of both species in 2010: while remarkably similar with around 14,500 genes each, *Volvox* has more genes that encode proteins for cell walls and the extracellular matrix – genes for sticking cells together.

Dependency and communication

In 1988, *Scientific American* published a seminal article by geneticist James Shapiro that challenged the view of microbes as single cells: 'Bacteria as Multicellular Organisms'. One of his examples was the cyanobacterium *Anabaena cylindrica*. Regular cyanobacteria photosynthesize and absorb atmospheric nitrogen at different times as metabolic reactions for the two processes interfere with one another. But the filamentous form of *Anabaena* consists of chains of cells that remain linked due to incomplete separation after division. These are specialized into photosynthetic cells, nitrogen-fixing heterocysts, akinetes (resting cells) and hormogonia that move around. The first two cannot reproduce, but the last two can, like the soma/germ division of a complex body.

> The principle of division of labour which appeared among multicellular organisms ... has gradually led to the production of greater and greater complexity in their structure.
>
> August Weismann

However, filamentous forms and other patterns are less manoeuvrable and increase competition for resources due to the high density of cells, so why form a body at all? In 2006, ecologists Gianluca Corno and Klaus Jürgens grew freshwater *Flectobacillus* along with the bacterivorous alga *Ochromonas*, and found that over 80 per cent of prey turned into inedible filaments formed from multiple elongated cells. So one trigger for the transition to multicellular life may simply be that a bigger body is less likely to get eaten by predators.

Individuality

Going from a group of cells to a multicellular body is a profound ecological change, from competition to cooperation, that redefines what it means to be an individual organism. During that transition, evolution by natural selection can occur on multiple levels based on whether group living affects the ability to survive and reproduce. If costs outweigh benefits, (as seen in bacterial biofilms), cell-level selection will quickly outpace group-level selection.

So how can multicellularity be stabilized? One scenario is that natural selection could favour a trait that benefits a cell's fitness in a group, but is costly if the cell leaves. In 2012, William Ratcliff tested this with experimental evolution in a normally unicellular yeast, *Saccharomyces cerevisiae*. He gave cells 45 minutes to settle in a test tube before transferring those at the bottom to a new tube, repeating this 60 times so artificial selection favoured heavy, multicellular clusters. Surprisingly, the yeast evolved a second trait besides sticking together: high rates of apoptosis – programmed cell death.

Based on mathematical models, Ratcliff and Eric Libby suggested that death of 'weak links' allows cells to overcome constraints in a test tube – breaking links produces smaller, fast-growing cells. Apoptosis is an adaptation to group living, but is maladaptive if cells break free of a cluster, as the high suicide rate makes them less competitive against other free-living cells. Traits like apoptosis could work as clicks on an evolutionary ratchet, entrenching cells in a group lifestyle and making it hard to revert back to a solitary existence.

The condensed idea
Cells lose individuality to gain specialized roles in a multicellular body

26 Circulation

Animals power most metabolic reactions by cellular respiration, which absorbs oxygen and nutrients, then excrete the waste products of metabolism. Within the body, substances move between cells and the environment through blood vessels or other transport networks – a circulatory system.

A three-dimensional body creates a physiological challenge, neatly illustrated by thinking about a sheet of bubble wrap: when flat, you can easily pop any bubble, regardless of the sheet's size, but if you roll the wrap into a cylinder, it is much harder to reach the central cells. In this metaphor, popping a bubble represents the rate of diffusion, the movement of molecules down a concentration gradient (high to low).

Diffusion is adequate for moving metabolites across the surface of a multicellular sheet, but not through a 3-D body. With more cells, volume increases faster than surface area, and it becomes impossible to meet the metabolic demands of cells by diffusion alone. One solution is to raise the area-to-volume ratio by folding: jellyfish do this by creating a body cavity that is not actually inside the body, but continuous with their watery environment. For a solid 3-D body, however, you need circulation to transport metabolites.

Open and closed systems

Circulation uses interconnected vessels – a 'vascular system' – and pumps that move blood through a body. In a closed circulatory system, blood stays in vessels while cells are completely bathed in 'interstitial fluid'. Metabolites are exchanged by diffusion across a layer called the endothelium, and interstitial fluid often drains into the lymphatic system, where it forms 'lymph' before being recycled back into blood. In an open system, blood empties into the body cavity or haemocoel, and vessels are not lined by an endothelium. Because there is no distinction between the three fluids, it is known as haemolymph or simply 'blood'.

All vertebrates use a closed circulatory system, whereas invertebrates use whatever suits their (more diverse) lifestyles: most molluscs, crustaceans and insects use open systems. Molluscs are interesting because, alongside bivalves like oysters and gastropods like snails, the group also includes cephalopods such as octopus and squid. These use a closed system with powerful hearts, an adaptation to an active lifestyle that includes swimming and predatory behaviour.

Cardiac and vascular systems

Today, we know that the heart of a vascular system is, well, the heart. But for almost 1,400 years, anatomy was dominated by the ideas of one man: the Greek physician Galen, born around 130 AD. Instead of circulation, Galen said that blood was produced by the liver using food from the gut, then consumed by tissues – an open system that did not recycle fluid. The liver was central to this system and blood was infused with 'vital spirits', a mix of air from the lungs and heat from the heart. The heart was not a pump, and the septum separating its left and right chambers included pores.

Galen's ideas went largely unchallenged until the 16th century, when Italian anatomists at the University of Padua highlighted his mistakes: in 1543 Andreas Vesalius showed that blood does not flow between the heart's two sides; in 1559 Realdo Colombo claimed it circulates via the lungs; and in 1603 Hieronymus Fabricius discovered that veins have one-way valves to prevent blood from flowing backwards. It was Fabricius' student, English physician William Harvey, who finally disproved Galen's dogma. Harvey measured the volume of blood that drained from various mammals and calculated that it was far more than could come from food. From this, he concluded that 'the blood in the animal body is impelled in a circle, and is in a state of ceaseless motion'. In 1616, Harvey started lectures at the College of Physicians to demonstrate his circulation theory in diverse animals, and also revealed the direction of blood flow by using a tourniquet on a human arm, causing vessels to become engorged and showing that arteries come from the heart while veins go towards it.

> The heart of animals is the foundation of their life, the sovereign of everything within them, the sun of their microcosm, that upon which all growth depends, from which all power proceeds.
> William Harvey

Harvey's 1628 book, *Anatomical Exercise on the Motion of the Heart and Blood in Animals*, includes a dedication to the king that states that 'all power proceeds' from the heart. But Harvey was only

Three-dimensional bodies originated twice: multicellular animals arose about 700 million years ago while plants first colonized terrestrial habitats 450 million years ago. Through convergent evolution, land plants found similar solutions to living with multiple, diverse cells, including a shoot-root axis analogous to animal head–tail morphology, and apical meristems – stem cells at growing tissue tips. Plants lack a circulatory system, but do have a vascular transport system that carries fluid through two types of vessels: phloem and xylem. Phloem tubes are filled with sap and lined with living cells that actively push sugars made by photosynthesis into the sap, from where they diffuse into cells. Xylem consists of dead cells that passively suck water and dissolved nutrients upwards. After diffusion across a membrane (osmosis) from soil into root, the liquid overcomes gravity through capillary action, and is used to replace water lost by transpiration out of leaf pores (stoma) or by evaporation from surfaces. The walls of plant cells contain cellulose and lignin that resist compression and other stresses, providing the structural support that lets plants grow tall.

half right: most animals have a single organ, but some have multiple hearts. Octopuses have one systemic heart for pumping blood to the body, plus two accessory hearts that supply the gills. Earthworms and other annelids do not have an organ, pushing blood through their closed circulation system by squeezing their body, a coordinated wave of 'peristalsis' or muscular contraction that resembles how food moves through your digestive system.

Circulatory systems are either single or double, depending on whether hearts are divided. The heart of a fish has one atrium that leads to one ventricle, so blood is pumped towards the body via their gas exchange organs – the gills. Birds and mammals have a double circulatory system where the heart is divided into left and right sides, separating deoxygenated blood directed to the lungs – a pulmonary circuit – from oxygenated blood destined for the body: the systemic circuit. Other vertebrates have partial separation into a left and right system.

Gas exchange systems

One part of the vascular system that Harvey failed to find were the blood vessels that supply cells. Although he suggested they exist, these capillaries were not observed until 1661, when Italian biologist Marcello Malpighi saw them while studying frog lungs under a microscope. Malpighi also proposed that the lung surface is where gases are exchanged between air and blood – something we now know occurs by diffusion across the capillary walls. Respiration uses oxygen and releases carbon dioxide, gases that are usually attached to respiratory pigments like the haem in haemoglobin, and often carried by blood cells. Malpighi discovered that insects do not transport gases in the blood, but use the tracheal system, which opens at pores in the exoskeleton (spiracles) and lead to a branching arrangement of tubes, bringing air close enough for diffusion into the blood that bathes cells.

Reptiles, mammals and birds also use branched tubes, with trachea and bronchi leading to inflatable air sacs, while fish force water across their gills and some amphibians rely purely on diffusion of gases across the skin. Physiological respiration is often described as a distinct process ending with gas exchange, but it is better to think of the respiratory and vascular systems as connected circulation.

The condensed idea
Transport systems overcome the drawbacks of diffusion

27 Ageing

Death is a natural phenomenon. In the wild, organisms usually succumb to environmental challenges such as predators, disease or accidental injury. Individuals who survive such 'extrinsic' causes of mortality then face 'intrinsic' mortality – they die of old age. Maximum lifespans differ between species, however, which raises the question of just why we age.

One outdated explanation for ageing – still heard today – is that individuals die to make room for the next generation. This implies that natural selection acts 'for the good of the group' to prevent overcrowding. As German biologist August Weismann said in 1889: 'Worn-out individuals are not only valueless to the species, but they are even harmful, for they take the place of those which are sound.' This naïve argument not only contributes to age discrimination in society, it also defies evolutionary logic because a 'suicidal' population is vulnerable to cheaters: if an immortal individual emerged, it would benefit from the sacrifice of others without incurring the cost of its own death. Its offspring could then spread their 'immortality gene' through a gene pool, eliminating ageing.

Evolutionary explanations

From both individual and 'selfish gene' perspectives (see chapter 46), immortality has a major advantage: an organism can continue to procreate. So how could ageing possibly persist? In a 1951 lecture, zoologist Peter Medawar offered two key insights. First, he distinguished between the ageing process and what he called 'senescence' – the biological symptoms that decrease body performance and increase the risk of extrinsic mortality from, say, a predator. Second, he pointed out that the strength of natural selection declines with age, so it cannot influence mutations that cause lethal conditions like cancer and cardiovascular disease.

Natural selection cannot notice a genetic mutation unless it creates a visible phenotype. A mutation that lowers performance in early life might cause an individual to fall victim to survival of the fittest, but a mutation that causes senescence later on is effectively invisible. This leads to the decline in the strength of selection over a lifetime: if a mutated gene reduces performance before you reproduce, it is not passed on, but if a mutation causes senescence after reproduction, it is too late – the genes have already been inherited. Causes of senescence can therefore accumulate over evolutionary time – Medawar's 'mutation accumulation' theory of ageing.

Body tissues are reproductive 'germ' or everything else: 'soma'. Germ transmits genes to the next generation, while soma is thrown away when organisms die. This is the basis of the 'disposable soma' theory proposed by Tom Kirkwood in 1977, which views ageing as a trade-off between two sides of evolutionary fitness: survival (growth, maintainance and repair of the soma), and reproduction (making germ cells like sperm and egg). Kirkwood's theory argues that since

Life extension

The most reliable way to extend lifespan is to eat less, known as calorie or dietary restriction, as demonstrated in diverse organisms. In the mouse, for instance, reducing food intake by 30–40 per cent promotes longevity, slows physiological signs of ageing and helps prevent disease.

Precisely how it works is not clear, but in 1999 molecular biologist Leonard Guarente suggested that proteins called sirtuins are involved. His team discovered Sir2, a sirtuin that extends lifespan of yeast cells, and a year later they found that it controls other proteins, resulting in altered metabolism and response to cellular stress. Animals have half a dozen Sir2 equivalents, including SIRT1, which geneticist David Sinclair has shown can be activated by small molecules to mimic calorie restriction, raising the possibility of developing drugs to achieve the same effects. One sirtuin-stimulating molecule is resveratrol, the chemical found in grape skin that (supposedly) makes drinking red wine good for your health. In 2006, two teams showed that mice given resveratrol can eat a high-calorie diet without gaining weight or developing diabetes. Some researchers dispute the effects of both resveratrol and sirtuin 'longevity genes', but the life-extending results of dietary restriction remain robust.

ecological resources like food are limited, so is the energy generated through metabolism. This leads to economic decisions when allocating resources between physiological processes: when times are tough, the priority is survival, but extra resources allows the luxury of reproduction.

Programmed lifespan

Life's instructions are encoded within genes, so is death also programmed by DNA? At first sight, it appears that way. In 1961, anatomist Leonard Hayflick showed that after growing cells in a Petri dish, they would stop after about 50 divisions, now called the 'Hayflick limit'. In the 1980s, molecular biologist Elizabeth Blackburn discovered that telomeres – DNA sequences that protect chromosome ends – are lost during cell division, which implies that telomere loss is a countdown timer to cellular retirement, which could explain the Hayflick limit. Animal studies have also found genes linked to longevity. In 1993, for example, biogerontologist Cynthia Kenyon identified a single mutation that doubles the lifespan of nematode worms.

But like many phenotypes determined by genes, senescence is also influenced by the environment. Female honeybees develop into queens or workers depending on the food they receive as larvae, but average life expectancy is two years for queens and months for workers – with no difference in DNA. In humans, the relative contributions of nature and nurture can be measured by comparing twins, who share almost identical genomes but rarely die at the same age: a 2004 survey of over 2,700 pairs of twins found that genetics explained just 20 per cent of age-related impairments, highlighting environmental impact on the rest.

Molecular mechanisms

Ageing of the body is driven by wear-and-tear to cells, such as an accumulation of proteins becoming misfolded due to stress and mutations in mitochondrial DNA. For example 'free radicals' (reactive forms of oxygen) are produced during respiration, and leak out of mutated mitochondria to react with molecules in the cytoplasm. Maintenance and repair systems help prevent cellular senescence, but their performance declines over time. In 1992, Alexander Bürkle measured the activity of one DNA repair enzyme (PARP1) in cells from a dozen mammals and found a relationship between enzyme activity and a species' maximum lifespan: at the extremes, humans have five times more DNA repair than rats, which only live for three to four years.

So why do we age? Maintaining and repairing cells uses energy and, according to the disposable soma theory, senescence is a legacy of trade-off in allocating resources between reproduction and survival. This helps explain why dietary restriction can extend lifespan, as survival becomes the priority when there is a lack of food. It also tallies with research by biogerontologist Linda Partridge, who has found that signals sent via insulin hormones and 'insulin-like growth factor' can sense nutrients and regulate processes like growth and metabolism. Organisms do not need to keep the body in perfect condition, just good enough to survive beyond the point of fertility, which determines their life history: in the wild, more than 90 per cent of mice die within a year, so their three-year lifespan leaves more than enough time to reproduce. Modern medicine and technology now defend humans against extrinsic causes of mortality such as disease and predators, so we die from intrinsic mortality. Humans worry about ageing simply because we live long enough to experience it.

> In the post-reproductive period of life, the direct influence of natural selection has been reduced to zero, and the principle causes of death today lie just beyond its grasp.
> Peter Medawar

The condensed idea
Lifespan is a trade-off between survival and reproduction

28 Stem cells

Early in development, animal cells have the potential to create almost any part of the body. Researchers hoping to harness that power for medical applications were once restricted to embryonic cells, but studies of mice and frogs reveal that stem cells can also be made by reprogramming specialized tissues.

Why are stem cells so special? During development, a fertilized egg divides and its descendants become specialized for different roles, from oxygen-carrying blood cells to protective skin. This process of 'differentiation' creates over 200 types of cell in the human body. The first to visualize how this happens was German naturalist and artist Ernst Haeckel. In 1868, he drew a 'tree of life' in which the central stem represented the ancestor of all life, a unicellular organism Haeckel called the *Stammzelle* – stem cell. In 1877, Haeckel extended the concept to embryology, proposing that a fertilized egg is also a stem cell. Differentiation forms a treelike hierarchy with the embryo as the trunk, specialized cells as leaves and stem cells as branches (but not twigs). Stem cell research owes a lot to one branch: haematopoiesis, or the formation of blood cells.

In 1896, German haematologist Artur Pappenheim described the progenitor of both red and white blood cells as a 'stem cell'. Then in 1905, he drew a genealogy of cells radiating from a central progenitor. The first big discovery in stem cell science came in 1960, when Canadian cancer researchers James Till and Ernest McCulloch found that some cells in the bone marrow of mice are sensitive to radiation. In 1963, the pair transplanted those cells into mouse spleens, where they multiplied to produce blood cells.

Potential

Till and McCulloch revealed the two key features of stem cells: they divide indefinitely, and have the potential to give rise to specialized cells. The power to produce other cell types depends on a cell's position in the differentiation tree. Haematopoietic stem cells are 'multipotent' because the branch makes different blood cells, while a fertilized egg is 'totipotent' as it creates the whole body. In most mammals, a hollow ball of cells called the blastocyst contains a 'pluripotent' inner cell mass that, if implanted, forms an embryo and all tissues except the placenta.

Embryonic stem cells were first isolated from mouse blastocysts by British embryologists Martin Evans and Matthew Kaufman in 1981. Human embryonic stem cells were grown in 1998 by American biologist James Thomson. Adult stem cells are rare and are only multipotent, and there is an alternative: pluripotent embryonic cells can be obtained relatively easily from 'spare' blastocysts discarded from in vitro fertilization treatment, but this raises ethical concerns, especially among people who believe life begins at conception, before the ball becomes an embryo.

Cloning

Another ethical issue for stem cells is growing a human through reproductive cloning. However, laws around the world currently prevent this and known research is on therapeutic cloning at the level of genes rather than individuals. Nevertheless, reproductive animal cloning has provided important insights.

Differentiation, for example, was once thought to be a one-way process: cells on the branch leading to skin could not backtrack and turn into blood. In the 1950s, experiments by Thomas King and Robert Briggs showed that when a frog nucleus is transplanted into a surrogate egg, animals developed normally, but transfer into a mature embryo meant fewer frogs. This suggested that something in the nucleus is lost during development. In 1962, British biologist John Gurdon came to a different conclusion. Studying African clawed frogs and using ultraviolet light to destroy DNA in a surrogate egg cell, Gurdon then used a micropipette to replace the nucleus with a mature one from the epithelium (lining) of a tadpole's gut. Of 726 eggs, most developed abnormally, but ten grew into tadpoles – the first animals cloned from non-embryonic cells. This not only proved that differentiation could be reversed, it suggested that the egg cytoplasm could effectively reprogram the nucleus.

Reprogramming

The various cells in your body have essentially the same set of genes, so what makes them different? Think of your genome as a computer's operating system. As you use it for specific tasks over time, you install software that causes it to become more specialized. The same is true in a cell, which contains proteins called 'transcription factors' that bind to switches on DNA and turn genes on and off. Transcription factors are epigenetic marks that get transmitted via the cytoplasm when a cell divides, programming its daughter cells to become a specific type. The egg's cytoplasm then wipes that software from the genome's hard drive.

The big breakthrough in genome reprogramming came in 2006 from Japanese scientist Shinya Yamanaka. Previous research had shown that stem cells activate transcription factors so Yamanaka created genetically engineered fibroblasts (specialized cells) where transcription factor genes were always active. One combination of four genes, now

> Moving the nucleus of a somatic cell into an egg, there is a remarkable re-programming effect ... changing it from the specialized type of the differentiated cell back into the stem cell type of an embryo.
>
> John Gurdon

called 'Yamanaka factors', produced cells with an appearance, behaviour and genetic activity similar to embryonic stem cells. Yamanaka's approach involves prompting, so it creates 'induced pluripotent' stem (iPS) cells. Researchers are unsure whether the four Yamanaka factors are best for making iPS cells – after a week of division, just one in a thousand becomes pluripotent, and how reprogramming works is still unclear.

Meanwhile, embryonic cells have been converted to eye cells to treat a common condition that leads to blindness: age-related macular degeneration. Stem cells therapy has also been achieved with iPS cells, which have the advantage of originating from a patient, minimizing risk of immune rejection. Fixing the body with your own cells is just around the corner.

The condensed idea
Programming DNA creates different types of cell

29 Fertilization

The fusion of sperm and egg deserves to be called the 'miracle of conception' – only one in a million human sperm get near the egg, for example. To beat these overwhelming odds, animals have strategies that help bring the two reproductive cells together.

When sexual reproduction combines sperm and egg from different individuals, the two gametes are often separated by tremendous distances. To get a crack at an egg, a human sperm must travel over a thousand times its length. The fertilization of an egg cell by a sperm is similar across the animal kingdom, as seen in 1875 by German embryologist Oscar Hertwig, who first described the fusion of male and female gametes of sea urchins, which have provided numerous insights into the fertilization process.

Laying eggs

Fertilization may be either internal or external, but both processes usually involve a liquid environment where sperm swim towards eggs. With external fertilization, mobile females can lay their eggs at a particular location – as in frogspawn – whereas sessile animals like corals release eggs into the water, where they either sink to the sea floor or riverbed, or spread further afield via broadcast spawning.

Internal fertilization involves sex organs: in mammals, the penis ejaculates semen into the vagina or uterus, and egg and sperm meet in the oviduct. The male gamete is smaller than the female gamete, which is why sperm come to eggs rather than the other way around. In popular culture, fertilization is often depicted as a massive number of sperm all racing towards one egg in a competition to fertilize it. In reality, however, sperm face a huge challenge in just finding the egg. Of the five million sperm ejaculated by mice, for instance, only about 20 ever reach the oviduct.

Guiding sperm

During external fertilization, sperm navigate using chemotaxis – movement towards the source of a chemical. This was described in the sea urchin *Arbacia punctulata* by American embryologist Frank Lillie in 1912. When Lillie added a drop of seawater – previously exposed to unfertilized eggs – to a suspension of sperm, the male gametes formed a ring around the egg extract, suggesting that female gametes secrete a substance to attract them. This chemical – resact – was isolated by pharmacologists J Randall Hansbrough and David Garbers in 1981. Resact opens channels in a sperm's membrane to allow ions to flow in and out of the cell, determining how frequently it beats its tail. In 2003, German biophysicist Ulrich Benjamin Kaupp showed that sperm can respond to a single resact molecule, suggesting that they count molecules over time to calculate a desired direction of travel.

For internal fertilization, at least in mammals, sperm are guided through rheotaxis – movement through a fluid. This was discovered in 2013 by Kiyoshi Miki and David Clapham, who saw that human and mouse sperm travel against the flow. Sexual intercourse stimulates the oviduct's walls to secrete fluids that push mucus and debris out of the way, clearing a path for sperm and offering them a cue for where to go. Rheotaxis drives natural selection between sperm, as only the strongest swimmers survive.

During internal fertilization, the closed space contains sperm from the same species – recognizing mates takes place before intercourse – so sperm make a beeline in the direction of the egg. With external fertilization, the open area contains other species, so swimming in circular loops maximizes the chances of finding an egg, while identifying specific chemicals helps prevent a sperm from trying to penetrate the wrong one. Chemotaxis also occurs in internal fertilization, but over short distances, with human sperm attracted to progesterone released near the egg.

In mammals, female gametes begin as 'oocytes' stored in the ovary, where the surrounding cells fatten them up with nutrients that feed a pre-embryo. Every menstrual cycle, a surge in gonadotrophin hormones prompts an oocyte to divide into unequal halves – a large egg and a small polar body. The unfertilized egg takes about 24 hours to mature and is released into the oviduct (Fallopian tube). Meanwhile, sperm consist of a head containing the nucleus and an acrosome at the tip, plus a tail (powered by mitochondria in the midpiece) that allows the male gamete to swim from the vagina or uterus to meet an egg in the oviduct. Sperm that successfully enter the oviduct are held in storage sites that allow females to release a few at a time. Conditions like alkaline pH (and progesterone in humans) ripen the sperm,

Test-tube babies

After British physiologist Robert Edwards failed to convince human sperm to fertilize eggs in a Petri dish, he teamed up with gynaecologist Patrick Steptoe, who used laparoscopic surgery to extract eggs from ovaries. The pair identified time of ovulation by monitoring menstrual cycles, then collected, fertilized and implanted an egg into a prospective mother's uterus for fertilization in vitro (Latin for 'in glass'). The first 'test-tube baby', Louise Brown, was born on 27 July 1978. More than 12 million babies have since been born through IVF, according to the International Committee for Monitoring Assisted Reproductive Technologies. One recent advance is Mitochondrial Replacement Therapy (MDT), where the nucleus from a mother's egg is transferred to an egg donated by another woman. This aims to stop a child from inheriting incurable diseases caused by genetic mutations in the energy-generating mitochondria in the cytoplasm of the mother's egg cell. Because the child carries genes from the nuclei of mother and father, plus DNA from the egg donor's mitochondria, they could be called a 'three-parent baby'. This is a bit misleading as nuclear DNA includes 20,000 genes and mitochondria usually only carry 37 – less than 0.2 per cent of a cell's total genome. As of April 2023, less than five children in the UK have been born through MDT.

giving it the capacity to penetrate the egg. 'Capacitated' sperm gain hyperactive motility, beating their tails in long, powerful strokes that propels them towards their final goal.

Fusing gametes

Gametes fuse after the sperm has crossed three barriers: a jelly layer, vitelline envelope and the egg cell membrane. In mammals, the jelly is an elastic matrix containing the cumulus cells that nurse an egg while it matures. Sperm break through this 'cumulus oophorous' layer using enzymes and brute force. The envelope or egg coat of mammals is called the 'zona pellucida' and contains various 'ZP glycoproteins'. When a sperm recognizes ZP glycoproteins, a capsule at its tip – the acrosome – releases enzymes

> As follows from numerous observations in both the animal and vegetable kingdoms, in the normal course of fecundation only a single spermatic filament penetrates into an egg.
> Oscar Hertwig

that carve a path through the zona pellucida. The acrosome reaction lets sperm reach the last barrier – the cell membrane – where surface proteins allow the egg and one lucky sperm to fuse. This triggers a change in the egg: it releases enzymes that chop up ZP glycoproteins to prevent other sperm from entering.

While waiting to be fertilized, the egg puts its cell division cycle on hold. During fertilization, the sperm delivers an enzyme that removes the block on the cell cycle: the egg finishes dividing, leaving a female pronucleus with half the normal number of chromosomes. This fuses with the male pronucleus, delivered by a sperm after penetration, forming a nucleus containing pairs of chromosomes.

The final stage of fertilization is a bit of a mystery. Both the sperm and egg are specialized cells that differ from other cells with the same DNA due to epigenetic marks that switch genes on/off. These marks should be wiped off to leave a blank slate, but without also erasing marks that parents deliberately added to their gametes. However this is achieved, the process leaves a fertilized egg, now a single-celled zygote, that develops into a complex multicellular organism like you – another miracle.

The condensed idea
The fusion of egg and sperm requires chemical cues

30 Embryogenesis

Physician William Harvey's 1651 book *On the Generation of Animals* opens with an inscription: *Ex ovo omnia* – 'All from the egg'. The statement went against Aristotle's idea that life arises by spontaneous generation from inanimate matter, and led developmental biologists to study how embryos are created.

Aristotle is best known as a Greek philosopher, but should also be considered as the first biologist. He made contributions to anatomy and embryology, working out the role of the placenta and umbilical cord during pregnancy, for instance, and stating that animals are born from eggs (oviparity), through live birth (viviparity) or after eggs hatch inside a mother's body (ovoviviparity, seen in sharks and certain reptiles). But while Aristotle believed in development from an egg, he also believed that animals arose from non-living matter such as mud.

Egg to embryo

For two thousand years there was little progress in embryology. With the invention of the microscope, scientists were able to describe early stages of development and studies of chick anatomy initially suggested that there was a 'preformation' of miniature adult organs in an embryo. Nonetheless, Harvey and Aristotle believed in 'epigenesis' (structures are created from scratch), a hypothesis that was proven true by German embryologist Caspar Friedrich Wolff, whose work on chicks showed that body parts are formed during development. In 1767, Wolff observed that the intestine begins as flat tissue that becomes folded – like the tube created when opposite edges of a sheet of paper are pushed towards each other on a desk.

Layers to organs

Modern embryology was founded by three friends, all from the Baltic region, born within a year of each other, who studied in northern Germany: Karl Ernst von Baer described the process of development; Martin Rathke compared similar structures across vertebrates; and Heinz Christian Pander discovered that organ systems originate from distinct embryonic layers. Pander spent just 15 months studying

Early embryos

Multicellular embryos during the early development of mammals and sea urchins. A 'morula' is any solid ball of cells created after cleavage of a fertilized egg (zygote), consisting of up to a dozen or so cells. A 'blastula' is a hollow sphere that often includes hundreds of cells and, in placental mammals, an 'inner cell mass' that forms the embryo. A 'gastrula' has two or three 'germ layers' that give rise to organ systems, where an animal starts to develop its body shape (morphology).

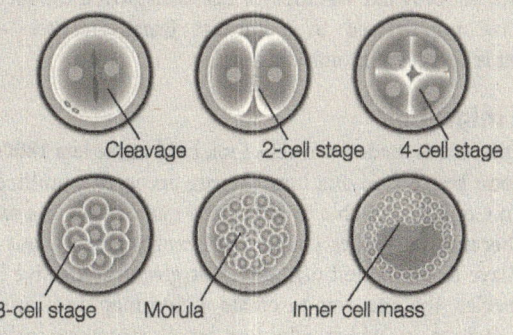

Cleavage   2-cell stage   4-cell stage

8-cell stage   Morula   Inner cell mass

chicks, during which he discovered that animal embryos have three 'germ layers': an ectoderm destined to form epidermis and nerves; an endoderm that forms internal structures such as the digestive system and organs like the lungs; and a mesoderm sandwiched between the other two, which makes blood and bones, heart and kidneys, gonads and connective tissues. Simple creatures such as sea sponges and jellyfish are 'diploblastic' because they have two layers (no mesoderm), while three-layered animals are 'triploblastic'. In 1817, Pander found that a layer only produces organs if the others are present. This principle of induction – cells prompting one another to develop – underlies body shape (morphology – see chapter 31) and the way that stem cells become specialized.

Von Baer studied chick development and discovered the notochord, a backbone-like rod that induces formation of nerve cells in the nearby ectoderm. In 1828, he described the difficulty of

> I have two small embryos preserved in alcohol, that I forgot to label ... They may be lizards, small birds, or even mammals.
> Karl Ernst von Baer

distinguishing between vertebrates, leading to 'von Baer's laws': first, general features look very similar in early embryos of different animals; second, specialized features develop from general ones (so feathers, hair and scales form from skin); third, an embryo from one species does not pass via developmental stages of another; and fourth, the embryo of a higher animal is never like a lower animal, but only like its embryo. These disproved the incorrect 'biogenic law' popularized by naturalist Ernst Haeckel – that development reflects evolutionary history.

Mapping migration

The theory that life is made from cells took hold in the late 1800s, and biologists soon began studying how the egg becomes a multicellular body. Edwin Conklin was able to follow the fate of cells in a sac-like sea squirt because its tissues contain different pigments, but other organisms have to be stained using dyes or given radioactive labels. For each species, the aim was to create a 'fate map', indicating the embryonic regions in which adult or larval structures originate.

Similar structures

While comparing vertebrates in the 1830s, German embryologist Martin Rathke described the pharyngeal arches, structures that develop into parts of the gill apparatus in fish, but the jaws and ears of mammals. In evolutionary terms, gills and ears are 'homologous' because they originate from a common ancestor. The most famous example of homology is forelimbs – a primate's arms, dolphin's flippers, bird's wings – which are the front legs in all tetrapods. Alternatively, similar forms can be 'analogous' because they do not develop from a shared ancestral structure, as in the wings of birds, bats and pterosaurs. Birds flap their 'arms', bats use fingers, while the extinct pterosaurs stretched a wing membrane from an extra-long fourth finger. Insect wings are not related to legs.

Certain cells migrate around the body during embryogenesis, as shown by American zoologist Mary Rawles in 1940 using cells from a region called the neural crest that later move into the epidermis. The primordial germ cells that become sperm or eggs also migrate, from yolk-rich cells to the gonads, while blood stem cells end up in liver and bone marrow.

Cell cleavage

Division of a fertilized egg to produce multiple cells begins along an axis dictated by distribution of yolk: the dilute end is the 'animal' pole, while the yolk-rich end is the 'vegetal' pole. Cleavage was observed in the late 1800s by Oscar Hertwig, who studied sea urchin cells. It was not until 1984, however, that Marc Kirschner identified the trigger: Maturation-Promoting Factor (MPF), a protein that tells cells to quickly flip between DNA replication and mitosis. This cleavage rate is faster than any other cell division: a fruit fly embryo grows 50,000 cells in 12 hours, for example. Until the 16-cell stage, any solid ball is called a 'morula', after which a central cavity appears, forming a hollow sphere or 'blastula'.

Gastrulation creates the three germ layers – ectoderm, mesoderm, endoderm – of the 'gastrula', leading to physical changes to the embryo. In sea urchins, for example, the endoderm is folded inwards and mesoderm cells migrate towards the sphere's centre. These processes involve coordinated growth and movement between neighbouring cells, a vital stage of embryogenesis. To quote developmental biologist Lewis Wolpert, 'It is not birth, marriage or death, but gastrulation which is truly the most important time in your life.'

The condensed idea
The creation of a body with distinct layers

31 Morphology

Whether an embryo becomes a limbless worm or winged bird, almost all animals have the same basic form, with three axes: front and back, head and tail, left and right. This essential structure is controlled using a common genetic toolkit and determined during morphogenesis, the creation of body shape.

Developmental biologists often name things after the effect of genetic mutations, which is why a protein that adds bristles to fruit fly larvae was named 'Hedgehog'. In the 1990s, when Harvard scientists found hedgehog proteins in vertebrates, they named them after different species. But Bob Riddle wanted a more exciting moniker for his molecule and convinced his boss Cliff Tabin to name it after a new videogame character from his daughter's magazine: Sonic the Hedgehog.

The Sonic hedgehog protein is a morphogen – a molecule that signals to cells, telling them to form a pattern. Such changes to shape are also influenced by direct cell-to-cell contact, which is how a receptor called Notch works. Such signals determine the fate of nearby cells, providing information about identity and position along each axis in three-dimensional space.

Back to front

The first evidence that cells determine the fate of their neighbours came from studying the dorso-ventral (back-front) axis. In 1924, Hilde Mangold took tissue from the 'dorsal lip' of a newt embryo and grafted it onto the ventral (stomach) side of another animal. Thanks to the different colours of the newts, the German embryologist could follow development at the gastrulation stage, when embryos have three layers (see chapter 30). The transplanted tissue seemed to 'organize' the outer ectoderm layer: cells destined to become epidermis turned into neural tissue, and a second dorso-ventral axis was created, producing twin, front-to-front tadpoles. Mangold tragically died before her study was published, but her supervisor Hans Spemann continued the work, winning a Nobel Prize for the 'organizer effect'.

For half a century, biologists thought the 'Spemann organizer' released a signal that encourages neural tissue to develop. But in 1989,

Horst Grunz and Lothar Tacke showed that when ectoderm from clawed frogs is split into cells, epidermis is only formed if the cells are reunited within an hour – any longer and they become neural tissue. So instead of prompting 'neuralation', the Spemann organizer *stops* cells from following their default fate.

A key molecule released by the organizer was revealed in 1996, a growth factor called 'bone morphogenetic protein 4' (BMP4). Morphogens like BMP4 are 'inhibitors' that prevent change, while 'inducers' such as Sonic hedgehog promote development. Sonic also helps determine the poles (dorsal or ventral) of the central nervous system. Philip Ingham showed in 1993 that Sonic is made by the notochord and base of the neural tube in zebrafish, while Andrew McMahon found the same in mice. The organizer effects of ectoderm and notochord, releasing BMP4 and Sonic hedgehog, respectively, show that a cell's fate depends on distance from the source of a morphogen. In 1969, the British developmental biologist Lewis Wolpert proposed that morphogens produce a chemical gradient across a tissue, forming boundaries that define cell identity. The back-to-front 'neurulation' by BMP4 and Sonic is one example.

Head to tail

The morphogen Bicoid, which creates the anterior-posterior (head-tail) axis in fruit flies, was discovered by Christiane Nüsslein-Volhard and Eric Wieschaus in 1980, along with a dozen other genes, including the original Hedgehog. Developmental biology owes a lot to flies: in 1915, Calvin Bridges, a student in Morgan's famous fly lab at Columbia University, discovered bithorax. When mutated, this 'gene' will double features in a fly's mid-section. In 1978, American geneticist Edward Lewis found that bithorax is actually a cluster of genes, and that the position of a mutation matches where it affects the body. From this, he reasoned that each body segment's identity is determined by the order in which genes are activated in the cluster – in response to a morphogen gradient along the head-tail axis. In 1984, teams led by Walter Gehring and Matthew Scott found that bithorax and another cluster include a genetic sequence called a 'Hox' that lets an encoded protein bind DNA. Hox proteins therefore control development by flipping DNA switches. In 1989, biologists discovered Hox in everything from frogs and fish to mice and men.

Left to right

From the exterior, most animals look identical on both sides, but bilateral symmetry does not apply to internal organs – your heart is one example. If organ orientation is flipped in humans, it causes a rare condition, 'situs inversus', that affects at least 1 in 25,000 people. In 1995, geneticist Cliff Tabin, boss of the Harvard lab that isolated Sonic hedgehog, noticed that the protein is briefly produced asymmetrically in a chick embryo, on the left side of a 'primitive node'. After Tabin's team observed a protein called Nodal being made on the left, they forced cells to make Sonic on the right, which sometimes inverted the heart.

> My original concept was to give them names of vertebrate hedgehog species ... this is before Sega had brought the computer game to the U.S. and I'd never heard of it, but Sonic hedgehog sounded good.
>
> Cliff Tabin

The 'Lefty' gene was discovered by Japanese biologist Chikara Meno in 1996, and later shown to be two elements: Lefty-1 and Lefty-2. The name is actually a misnomer because Lefty-1 helps cells maintain 'righty' identity, stopping Nodal from crossing a mouse embryo's midline to give cells a 'lefty' signal. In 1998, Nobutaka Hirokawa and colleagues found the ultimate reason for the left–right axis, showing that the primitive node has cells with hairlike cilia rotating in a leftward (anticlockwise) direction.

Limbs grow outwards from an embryo's ectoderm layer along the proximo-distal (near-far) axis. In 1948, American biologist John Saunders discovered an organizer at the tip of a budding wing in chicks, and showed that removing that tip – the apical ectodermal ridge (AER) – caused the limb to be truncated. The AER can also organize other embryonic layers: in 1957, Saunders showed that when mesoderm that is destined to become thigh is grafted onto the AER, it develops into foot structures. Saunders also found an organizer at the posterior (tail) end of the wing bud that, when transplanted to the anterior end, produced a mirror-image of duplicated digits. In 1975, Lewis Wolpert proposed that this zone of polarizing activity (ZPA) releases a morphogen, and in 1993 that molecule was revealed by Bob Riddle and Cliff Tabin. The morphogen was, of course, Sonic hedgehog protein.

The condensed idea
Molecular gradients
determine body shape

32 Colouration

From a zebra's stripes to colour-changing chameleons, animals display a dazzling variety of visual patterns. This colouration produces features necessary for survival and reproduction, such as camouflage to deceive predators or prey, and sexy signals to attract a mate. How does the body create these colour patterns?

Colours result from natural selection: very generally, uniform colour is an adaptation to the physical environment – skin tone to protect against sunlight or absorb heat, for example – while the evolution of patterns is driven by biological interactions. As the only organisms with eyes, animals are the target audience for nature's visual displays, which have two purposes: crypsis (camouflage and other ecological strategies that make individuals more or less conspicuous) or communication (which sends honest signals, as in the warning colouration of poison dart frogs, or deceptive, as in mimicry). As discussed by Darwin in his 1871 book *The Descent of Man and Selection in Relation to Sex*, colourful signals enable females to choose mates.

Pigments and structures

Mammals are shades of grey or brown because they can only make variations on a single pigment – melanin – which is added to skin and fur. Pigments are contained within lipid bubbles inside 'chromatophores', specialized cells that create shades and colours through the distribution of the pigment-containing bubbles. Other vertebrates gain colours from different sources, such as the yellows and reds of carotenoid molecules in food – creating pink flamingoes, for instance. Birds and mammals have one type of chromatophore cell – branching 'melanophores' – while fish have several types.

Colour is determined by the wavelength of light reflected into the eye by an object. The blue-on-brown 'eyes' of a peacock's tail have one underlying pigment, but a feather's structure scatters light to create two colours. This physical effect was first reported in the 1665 book *Micrographia* by English polymath Robert Hooke, who saw that feathers have thin plates that resemble mother-of-pearl shells. Structural colouration also creates the silvery-blue colour of fish – the membranes of iridophore cells contain stacks of reflective plates made from guanine, one of DNA's chemical letters. In fact, almost all blue and iridescent colour, from butterfly wings to bird plumage, is due to microscopic or nanoscopic structures that bounce light.

Patterns

Dorso-ventral countershading, where the back is darker than the stomach, is a simple yet effective pattern that offers basic camouflage to animals such as fish and birds. From above, it is more difficult to see them on the ground or in deep water. From below, their lighter underside is harder to spot against a bright sky. Development of countershading is controlled via hormones, such as the 'agouti-signalling peptide' in mammals, discovered in 1994 by American researchers led by Richard Woychik, William Wilkison and Roger Cone.

How are complex patterns created? Chromatophores originate from the neural crest, a transient collection of embryonic stem cells that also gives rise to parts of the head and peripheral nervous system. As an organism develops into an adult, its chromatophore cells divide and migrate, creating different colour intensities and patterns. The zebrafish *Danio rerio* is the model organism for studying this in vertebrates, and researchers have identified over 100 genes that influence pattern formation. In 2003, for example, German biologist Christiane Nüsslein-Volhard showed that mutations in the appropriated named 'leopard' gene altered the interactions between different chromatophore cells, causing animals to develop spots instead of stripes. Zebrafish development has thus helped explain how the leopard got its spots.

> The males are almost always the wooers ... with the most brilliant or conspicuous colours, often arranged in elegant patterns, whilst the females are left unadorned.
> Charles Darwin

Dynamic change

While chameleons can use camouflage, they mainly change colour for communication. Lizards similarly turn from green to red when excited. In many animals with dynamic colour change, it is assumed that this is controlled by physiological colour change, by hormones, which signals a chromatophore cell to disperse or aggregate its pigment-containing bubbles, altering brightness. But in 2015, biologist Michel Milinkovitch revealed that panther chameleons change colour using crystals in a layer of iridophore cells in the skin. These 'photonic nanocrystals' are made from guanine, like the reflective stacks in fish cells. Chameleons can 'tune' the spacing between atoms in the crystal lattice to adjust the reflected light. As well as camouflage and courtship display, this might also enable thermoregulation.

Bioluminescence and fluorescence

While the colours of most organisms depend on reflected light, some can generate their own illumination. The light is emitted through chemical reactions – the pigment luciferin glows after being combined with oxygen, while aequorin is activated with calcium. Bioluminescence is especially common in invertebrates, and plays the same role as colour patterns, for communication or crypsis: a glow-worm warns predators of toxicity and adult fireflies aim to attract a mate, for example, while the firefly squid emits light for camouflage that works like countershading.

A light-emitting animal has also provided one of the most useful tools in molecular biology: GFP (green fluorescent protein). As with other fluorescent molecules, GFP glows when activated by light, including that produced by aequorin. Both molecules were purified from a jellyfish in 1961 by Japanese scientist Osamu Shimomura. In 1992, Douglas Prasher isolated the GFP gene and proposed using it as a 'reporter' for visualizing genetic activity: insert it next to a gene of interest and you can see if both are switched on, based on the glow of GFP. Two years later, another American biologist, Martin Chalfie, demonstrated this in nematode worms. Roger Tsien later created other colours, kickstarting a rainbow revolution that has shed light on biological processes.

The masters of colour change are not chameleons, but cephalopods – octopus, cuttlefish and squid. A sophisticated nervous system and eyes allows cephalopods to rapidly assess a visual scene and match their body to the background using light and dark patterns that disrupt the animal's recognizable outline and shape. Unlike vertebrates, cephalopod chromatophores are not really cells, but organs – elastic sacs containing pigment and muscles, controlled by motor neurones. So instead of relying on the relatively slow action of hormones, their chromatophores are directly controlled by the brain, enabling quick colour change. *Octopus vulgaris* can go from fully camouflaged to conspicuous in two seconds.

Curiously, although matching the background requires being able to see it, most cephalopods are actually colour blind. This was proven in 1973 by British zoologist John Messenger, whose trained octopuses could discriminate between brightness but not hue (colour). And in 2005, biologists Lydia Mäthger and Roger Hanlon put cuttlefish on chequerboards of yellow/blue or green/grey squares, which matches the peak wavelength (492nm) of its visual pigment. The cuttlefish failed both tests, responding with a uniform body pattern. And yet despite their colour-blind camouflage, cephalopods are still able to fool other animals.

The condensed idea
Animal colour patterns are not always as they appear

33 Immunity

O rganisms are under constant attack from pathogens and parasites that aim to reproduce by exploiting a host's internal resources, causing infections that lead to disease and sometimes death. As a result, living things have developed powerful defence systems to fight off foreign invaders.

The ability to resist infection was first observed in humans by Greek historian Thucydides, who noted that people who survived the Plague of Athens in 430 BC were then protected from disease. Immunity became a science in 1796, when British physician Edward Jenner showed that exposing patients to cowpox virus provided resistance to smallpox. Over the next two centuries, researchers revealed that most organisms have an immune system with two branches: innate and adaptive.

Innate immunity

Physical barriers are a built-in first line of defence against invasion. Microbes have a cell wall, while multicellular organisms are covered in largely impenetrable layers – a plant's cuticle, an insect's exoskeleton or a vertebrate's skin. Vulnerable regions secrete mucus to trap potential invaders, areas that cover the body's entire surface or weak spots at the openings – amphibians are slimy all over, for instance, whereas only your nostrils ooze snot. Mucus contains protective molecules such as 'defensins', small proteins that work by punching holes in membranes, causing an invader's cells to spring a leak. As a result, the majority of pathogens simply fail to break into the body.

However, invaders that succeed must then avoid a host's surveillance system, which senses intruders via molecules characteristic of microbes. The sugary polymer lipopolysaccharide is unique to bacterial cell walls, for example, and such 'pathogen-associated molecular patterns' (PAMPs) match receptors on host cell surfaces. In animals, PAMPs can be detected by the complement system, a set of circulating proteins discovered by Belgian microbiologist Jules Bordet in 1895. Complement can combine into a 'membrane attack complex' that works like defensins, bursting invaders, or stick to PAMPs like a mark of death. In 1882, Russian biologist Ilya Metchnikoff observed that foreign material is surrounded and swallowed by phagocytes. These immune cells devour anything coated in PAMPs, complement proteins or antibodies, then digest them with acid, enzymes and free radicals.

Immune cells can serve as security guards, patrolling the circulation to check identity cards – the collection of molecules on a cell's surface – as viruses and other intracellular parasites often produce suspicious molecules that reveal their presence. In vertebrates, internal molecules are presented for inspection at a cell's surface by MHC (major histocompatibility complex) proteins. These vary in the ability to match molecules, so an individual's MHC genes influence their natural immunity to a particular pathogen. When an invasion is detected, both immune and infected cells release chemical signals that sound the alarm that the body is under attack, recruiting further help to the site of infection. Cells can also tell nearby blood vessels to dilate for easier access, leading to the accumulation of defences that causes the swelling and redness of inflammation.

Viruses infect all forms of cellular life, so all organisms have dedicated antiviral defences. Bacteria and archaea use 'restriction enzymes' that cut genetic material at recognizable sequences. Cells have double-stranded DNA genomes and produce single-stranded RNA transcripts, so the presence of long double-stranded RNA

HIV

Retroviruses insert their genes into a host's DNA, replicate quickly and evolve 1000 times faster than hosts. The best known is Human Immunodeficiency Virus. The HIV viral structure consists of two RNA strands surrounded by a conical capsid of proteins and a spherical outer envelope of fatty phospholipids, into which about 70 HIV proteins are embedded. These allow the virus to attach to, invade and replicate inside the body's white blood cells. Breakdown of these cells, which protect the body against pathogens, leads to Acquired Immunodeficiency Syndrome (AIDS). Since its emergence in the 1980s, HIV/AIDS is a global pandemic, with about 39 million people infected worldwide.

Immunity is often described in terms of friend and foe, 'self' and 'non-self'. But this ignores threats from within (cancer is self, and far from friendly) so it is better to use words like foreign and familiar: foreign antigens include viral RNA and pathogen-associated molecular patterns recognized by the innate immune system, whereas antigens become familiar through 'immunological tolerance', which is behind the ability to distinguish self from non-self. It occurs naturally during early development, and trains the body's defences to be unresponsive to its own antigens. If it fails, the body will attack itself and cause autoimmune disease. If an immune system is not trained to ignore harmless substances such as pollen or food allergens like peanut proteins, an allergy results.

(dsRNA) – molecules created by many viruses while replicating their genome – is a telltale sign of infection. In 1998, American biologists Andrew Fire and Craig Mello revealed that nematode worms specifically chop up dsRNA. Studies in other species showed this involves an 'RNA-induced silencing complex' of enzymes with names such as Slicer and Dicer. Many animals and plants use this process of RNA interference, while vertebrates produce 'interferons' – signalling molecules that recruit immune cells and tell neighbours to slow metabolic activity, disrupting viral processes.

Adaptive immunity

During an organism's lifetime, its immune system will adapt to recognize and remember invaders. In 2005, microbiologists found that bacteria and archaea store genetic material from viruses in their DNA: Cas enzymes chop up a viral genome, then paste bits into the host's genome as 'CRISPR' sequences – genetic memories of the viral invader that can be used as templates to target and cut matching genes if a virus reinvades. Prokaryotes were once considered too simple to have 'immunological memory', but given the existence of the CRISPR/Cas system, it is possible all life has some kind of adaptive immunity.

Mammals have the most sophisticated adaptive system, thanks to antibodies that adapt to match unfamiliar antigens ('antibody generators'). They are made by two types of immune cells that patrol the blood and lymphatic systems: B-cells, which release antibodies into the circulation, and T-cells with antibody-like surface receptors. Antibodies adapt to antigens through a process like Darwinian evolution. In 1974, Japanese biologist Susumu Tonegawa discovered that parts of antibody genes were mixed and matched, like the crossover between chromosomes when producing sperm or egg. When a B-cell encounters a foreign antigen, it divides rapidly, introducing errors during DNA replication. These two processes – somatic recombination

> Morbid matter of various kinds, when absorbed into the system, may produce effects in some degree similar.
> Edward Jenner

and hypermutation – create a huge variety of antibody proteins. As Australian virologist Frank Macfarlane Burnet proposed in 1957, immune cells then pass through a filter similar to natural selection: the fittest cells, whose antibodies match the antigen, survive and multiply, eventually overwhelming the pathogen. Antibodies can neutralize a pathogen's invasive ability, and also tag invaders for detection by parts of the innate immune system. Once an invader has been defeated, a few B-cells will continue to circulate as 'memory cells'. Adaptive immunity can take days or weeks to develop, but it provides long-term protection. Such immunity can be acquired artificially: modern vaccines contain a defective pathogen (or its antigens) that the adaptive system can use to make antibodies that would match a genuine invader.

The condensed idea
Defences are innate or adapt to foreign invaders

34 Homeostasis

n order to survive, an organism needs to keep conditions inside the body relatively stable. This occurs through homeostasis, a system of physiological processes that continually compensates for changes in the outside world that might otherwise damage the internal environment.

For two thousand years, Western medicine was based on the concept of humours: four fluids – blood, phlegm, yellow and black bile – thought to cause illness when out of balance. The idea was promoted by Hippocrates (460–370 BC), the great Greek physician who insisted that diseases have natural rather than supernatural causes. Support for the idea of humours dwindled in the 19th century with the development of immunology, and scientists switched to debating whether illness is mainly caused by germs, or the body's response to threats – what Canadian doctor William Osler compared to 'seed and soil'. Louis Pasteur, who founded the germ theory of disease, believed in seed, while his compatriot, physiologist Claude Bernard, was on the side of soil – the body's immune response. According to legend, Pasteur eventually conceded, saying: 'Bernard was right. The germ is nothing, the soil is everything.' •

Bernard discovered several ways the 'soil' responds to changes. Based on his experiments, mainly in mammals, Bernard developed the concept of the 'milieu intérieur' – the internal environment, a fluid that bathes and nourishes cells. Around 1876, he made a more profound suggestion: animals and plants keep their inner world constant through physiological mechanisms that continually compensate for external changes. In 1929, American physiologist Walter Cannon named this 'homeostasis'.

Negative feedback

Cannon's research in the early 20th century extended Bernard's theory. He suggested that organisms try to keep variables in the internal environment near an ideal value, such as 37°C (98.6°F) for human core body temperature: 'Automatic adjustments within the system are brought into action, and thereby wide oscillations are prevented and the internal conditions are held fairly constant.' Today, we might call those automatic adjustments a negative feedback loop.

Homeostasis treats the body like a kind of machine, a cybernetic system. Each physiological variable is regulated by a 'homeostat' that works like the thermostat in your home, through negative feedback: a thermostat turns the heating system on when the house is cold, and off when it gets warm (perhaps even activating an air-conditioning unit instead). In mammals, a temperature rise prompts sweating and diverts blood flow towards skin, helping heat loss, while a drop triggers shivering to generate heat through muscle activity, and constricts surface vessels to redirect blood to internal organs. Similar negative feedback loops control other variables, buffering the body against external fluctuations.

Acute and chronic stress

We have Hans Selye to blame for 'stress'. In 1936, the Austrian-Canadian physiologist described experiments where he subjected rats to a variety of 'noxious agents', including cold exposure, excessive exercise and formaldehyde injection. Regardless of the stressor, he saw that the same pathological symptoms would appear, a syndrome of enlarged adrenal glands, wasting of immune tissues and bleeding gut ulcers. Selye used 'stress' to describe this response, which redefined the word from meaning a physical force that produces strain or deformation, to a biological force that opposes change, effectively returning the body to a state of homeostasis.

> All the vital mechanisms ... always have one purpose, that of maintaining the integrity of the conditions of life within the internal environment.
> Claude Bernard

Selye called the response a 'General Adaptation Syndrome' and proposed three stages: an initial 'alarm reaction', adaptation through 'resistance' and 'exhaustion' that can lead to death. This is now integrated into the modern hypothalamic-pituitary-adrenocortical axis, or HPA axis.

In 1976, Selye defined stress as 'the non-specific response of the body to any demand'. 'Non-specific' relates to his experiments that suggested rats respond to a stressor with the same symptoms. This led to the 'stress syndrome' concept that, like 'adrenaline rush', is outdated. In 1998, for example, David Goldstein and colleagues tested Selye's

During an emergency, organisms prioritize short-term survival. In vertebrates, the brain activates the Hypothalamic-Pituitary-Adrenal (HPA) axis, a negative feedback system involving three endocrine (hormone) glands: the hypothalamus and pituitary gland above the brainstem, and adrenal glands above the kidneys (or interrenal organ in amphibians and fish). After exposure to a stressor, the sympathetic nervous system triggers an immediate release of catecholamines – adrenaline and noradrenaline – from the adrenal glands, shifting the body into the so-called 'fight or flight' mode so muscles and metabolism are ready for action. The hypothalamus then secretes corticotropin-releasing factor (CRH), a neurotransmitter that prompts the pituitary to release adrenocorticotrophin (ACTH), which in turn circulates through the bloodstream and stimulates the adrenal glands to produce glucocorticoid (GC) hormones – cortisol in most mammals and fish, corticosterone in birds and reptiles. The surge in GC hormones in the blood, a signature of an acute stress response, peaks within half an hour, altering an organism's physiology and behaviour so it can either cope or move away from the stressful situation. The negative feedback loop kicks in as GC hormones are detected by the brain, shutting down the HPA axis, which gradually returns an organism to its pre-emergency state.

theory by subjecting rats to stressors – such as cold and formaldehyde – and found different hormone levels in the HPA axis. Rather than a single, non-specific 'stress response', stressors will prompt different responses, each with a characteristic physiological signature.

Allostasis

Resting skeletal muscle in the human body uses about one litre of oxygenated blood per minute, but peak effort requires nearly 20 times more. Physiological activity is dictated by the demands of the outside world. In 1988, American neurobiologists Peter Sterling and Joseph

Eyer suggested that organisms can maintain stability through change, and introduced the term 'allostasis', meaning 'other sameness'. Using the home thermostat analogy, a desired temperature is set for morning and evening, summer and winter. Settings for each variable – like core temperature and blood glucose – are coordinated by the brain. Stressors are the equivalent of leaving a door or window open, which creates an 'allostatic load' that puts the body under strain, causing wear-and-tear to its equipment, which manifest as chronic stress.

The original theory of homeostasis, that an organism maintains its inner environment around constant states, is far too simple. Human physiologists prefer allostasis, but many biologists now combine the basic ideas of Bernard's milieu intérieur, Cannon's negative feedback and Selye's stress responses. Rather than the mechanisms that prevent internal change, homeostasis is now a holistic concept based on surviving environmental challenges.

The condensed idea
Compensating for external changes by adjusting internal conditions

35 Stress

Pushed out of its comfort zone, an organism's inner environment can become unstable, leading to stress – a loss of homeostasis. Stressors can include heat and cold, famine and drought, physical and psychological injury: all can trigger physiological responses that are not always useful.

How do organisms respond to stressful conditions? One strategy is movement: the sensitive plant *Mimosa pudica* folds its thin leaves when touched, for example, while many animals can curl up into a ball to shield themselves from the outside world – as in woodlice, armadillos and humans with a hangover. Running away is another option, but trees cannot uproot themselves and snails go nowhere fast. Rapid environmental changes cause harm regardless of whether an organism can move, however, which is why life has evolved similar mechanisms to cope with sudden stress.

Heat shock

One day in 1962, Italian geneticist Ferruccio Ritossa looked at his fruit fly chromosomes under the microscope and noticed some parts looking a bit puffy. Checking the incubator he used to grow cells, he realized that a colleague had accidentally raised its temperature. Chromosomes were normal at 25°C (77°F), but when Ritossa deliberately incubated cells for 30 minutes at 30°C (86°F), puffiness appeared. This was a year after Sydney Brenner showed that DNA's genetic information is carried to the protein-making ribosomes by an intermediate molecule – messenger RNA – and Ritossa thought his chromosomes puffed up because genetic activity was making RNA, but the encoded protein was not isolated until 1975. As it is made in response to temperature-induced stress, it was named heat shock protein 70 (HSP70).

Heat shock proteins are vital to cells. Metabolic processes are affected by temperature, and the folding of the proteins that control them can be altered by heat. Misfolded proteins can interact and aggregate in abnormal ways, as in neurodegenerative diseases like Alzheimer's. In 1984, British biologist Hugh Pelham showed that heat triggered protein aggregation in mouse and monkey cells, but extra HSP70 allowed cells to recover more quickly. Pelham later proposed that HSP70 works by binding to abnormal structures, prising them apart to release correctly folded proteins. There are numerous heat shock proteins, all belonging to a group of 'molecular chaperones' that guide protein folding.

Cellular stress is detected by 'heat shock factors' that attach to DNA switches and activate genes. Those genes encode heat shock proteins after being transcribed to RNA (causing the chromosome puffs). Heat shock factors are held back from DNA by other proteins, but heat changes their shape, prompting them to let go. The factors are found across life, from complex cells to bacteria. Moreover, heat is not the only stressor that triggers this response.

Extremophiles

Some microbes thrive in conditions that most other organisms would find stressful – or even lethal. These extremophile bacteria and archaea include heat-loving thermophiles that can reproduce at up to 121°C (250°F), halophiles that like high salt levels and acidophiles that prefer pH less than 3. Researchers have exploited these abilities in biotechnology, the best example being *Thermus aquaticus*, a bacterium isolated from a geyser in Yellowstone National Park in 1969: one of its enzymes, Taq polymerase, was later isolated and is now used to copy DNA in labs around the world.

The most impressive extremophile is a microbe so tough that it has been nicknamed 'Conan the bacterium'. *Deinococcus radiodurans* was discovered in 1956 after scientists tried to sterilize cans of corned beef by zapping them with ionizing radiation. In 2006, Croatian-French biologist Miroslav Radman found that shattering the organism's genome triggers a special DNA repair mechanism, using undamaged genetic material as a guide for stitching its chromosome back together.

Hormone systems

Physiological stress is a multicellular organism's response to an adverse environment, but stress comes from within, too, as when the human brain experiences anxiety. Certain cells are especially sensitive to stressors – for example, nematode worms have two temperature-sensitive nerve cells that prompt the release of hormones to broadcast stress signals around the body.

Most vertebrates control stress responses through hormones from two sides of a stress axis: catecholamines such as adrenaline, which maximize the chances of short-term survival, and steroids called glucocorticoids, which prepare the whole body for adverse conditions in the long term. Many invertebrates use a similar system, albeit with different hormones. Plants are far more pessimistic: their cells are geared to expect the worst unless they detect growth hormones called gibberellins.

> Anything that causes stress endangers life, unless it is met by adequate adaptive responses; conversely, anything that endangers life causes stress and adaptive responses.
>
> Hans Selye

The power of glucocorticoids has been harnessed in medicine. In 1936, American biochemist Edward Kendall isolated six steroid compounds from cow adrenal glands that improved muscle strength in dogs and rats. Kendall's Compound E was especially effective, but he could only obtain small amounts. However, industrial chemist Louis Sarett synthesized Compound E artificially, and by 1948 it was available for human trials. Physician Philip Hench of the Mayo Clinic gave it to a female patient crippled with rheumatoid arthritis – within a week, she walked out of hospital and went on a shopping trip.

Renamed cortisone, Kendall's Compound E was hailed as a wonder drug for its anti-inflammatory properties, despite side-effects like water retention and psychosis. In 1985, a team led by geneticist Ronald Evans identified a glucocorticoid receptor found inside almost all vertebrate cells. It was later discovered that cortisone is an inactive molecule, but the body converts it to a similar steroid, 'cortisol'. Once cortisol enters a cell, it binds to its receptor, ultimately changing the activity of thousands of its genes.

Natural stress

Hormone expert Hans Selye defined stress as 'the non-specific response of the body to any demand'. After studying the effects of various noxious agents on lab rats in 1936, he suggested that chronic activation of stress leads to 'General Adaptation Syndrome': various stressors always cause similar symptoms. This would tally with the effects of cortisol, which suspends repair so wear-and-tear accumulates. Studies show that different stressors have different pathology, however.

As a concept, stress was developed from studying animals in artificial environments, but while long-term stress kills humans and captive species, this may not occur in natural populations. Wild animals must regularly face chronic stress, and yet many do not just die from a stress response, but from starvation. In 2013, ecological physiologist Rudy Boonstra reviewed several cases of predation, one of nature's most stressful interactions, comparing levels of glucocorticoids produced by prey. Lemmings and voles did not seem too worried about weasels, the elk-wolf relationship was also relatively stress-free, whereas snowshoe hares and forest squirrels live in fear of predators. Boonstra believes this reflects differences in 'life history' – natural selection has driven some species to get used to stressors, so their hormones do not go haywire.

The condensed idea
Physiological stress responses can help or harm the body

36 Body clocks

The old proverb 'the early bird catches the worm' is relevant to many living things. This is not because being a 'morning person' is a more successful life strategy, but because keeping track of time allows an organism to match its physiology and behaviour to predictable changes in its environment.

Survival depends not only on adapting to environmental change, but also anticipating the day ahead: flowering plants get ready for pollination by opening their petals at the right time; nocturnal mammals prepare to forage by waking up before sunset. The first to study this phenomenon was French geophysicist Jean-Jacques d'Ortous de Mairan in 1729. After noticing that leaves of *Mimosa pudica* opened in the day and closed at night, he placed a plant inside a cupboard to test whether it was responding to sunlight. Despite the darkness, the leaves continued to move in a rhythmic manner. De Mairan believed his plants were reacting to other external cues, such as temperature or magnetic fields, rather than keeping time.

Circadian rhythms

Daily behaviour often reflects physiological processes that follow an oscillating pattern that roughly matches the 24-hour cycle of Earth's rotation. Levels of the sleep hormone melatonin rise at night and fall during the day, while body temperature does the opposite. In the 1930s, German botanist Erwin Bünning discovered that varieties of bean plant took different periods of time to cycle through leaf movements, and when two varieties were crossed, the hybrids had periods of intermediate length. This suggested an internal timekeeper, a biological clock with a daily circadian rhythm.

Such rhythms are built upon a natural cycle called a free-running period, which is not exactly 24 hours (hence the term 'circadian' – from *circa diem*, Latin for 'around a day'). Sleep researcher Charles Czeisler has found the average free-running period to be 24 hours and 11 minutes in humans. The period can be measured via physiological markers such as hormones or through behavioural rhythms, like the daily activity of rodents on a running wheel or the peak time at which adult fruit flies emerge from their pupal stage.

Keeping time

The modern science of chronobiology began in an outhouse in the Rocky Mountains. In 1952, British researcher Colin Pittendrigh of Princeton University spent the summer at a field station in Colorado and decided to repeat one of Bünning's experiments. The German had found that after placing flies in constant darkness and dropping the temperature from 26 to 16°C, their free-running period – based on emergence as adults – was delayed by 12 hours. Since metabolism consists of chemical reactions that are influenced by temperature, warm surroundings would prompt circadian rhythms to speed up, cold would cause them to slow down, making the biological clock a useless timekeeper.

Pittendrigh created one darkroom near a pressure cooker, another in an outhouse near an abandoned mineshaft. Unlike Bünning, he discovered that the peak in fly emergence came only one hour later in the cold outhouse. When he did the experiment again at Princeton, he got the same result. In the late 1950s, Pittendrigh and others showed this kind of temperature compensation occurred in various organisms, including unicellular protists and spore moulds, suggesting the biological clock is ubiquitous across life.

Staying in sync

Although a good clock is insensitive to environmental fluctuations such as temperature, it must also be possible to adjust the time. This is done using an external cue, what the German Jürgen Aschoff called a *Zeitgeber* (German for 'time-giver'), that synchronizes the internal clock with the outside world in a process called 'entrainment'. Zeitgebers stop the free-running period from falling out of sync with Earth's 24-hour cycle.

The key zeitgeber for most organisms is light. The 'master clock' or pacemaker in mammals is the SCN (suprachiasmatic nucleus), a cluster of nerve cells in the hypothalamus, above the brain's central chasm. In 1972, American neuroscientists Robert Moore and Irving Zucker independently discovered that damaging the SCN caused rats to lose circadian rhythms. Moore detected abnormal adrenal corticosterone levels, a physiological response to stress, whereas Zucker saw changes in drinking and movement behaviours, with animals being active at unusual times.

Sleep is the most important state regulated by body clocks. In humans, the dark-light and sleep-wake cycles have been roughly in sync for millennia, but technology has changed that: air travel causes jet lag because sunlight clashes with a clock saying 'time to sleep', while artificial light allows people to regularly reset their clocks, as occurs during shift work. Dark-light and sleep-wake cycles are controlled by a master clock and sleep homeostat that are both located in the hypothalamus. The link between tiredness and hunger is not surprising given that the hypothalamus controls both behaviours, and explains why we crave a midnight snack when we should be sleeping. Besides the brain's master clock, there are 'slave clocks' dotted around the body that are not synchronized by light, but by other external cues. The liver adjusts your internal time whenever you eat, for instance. Inappropriate syncing at night leads to poor quality, fragmented sleep patterns and health problems such as depression and obesity.

The master clock is reset during a critical period at dawn or dusk. In mammals, light is detected by the eye, using a unique type of cell not used in normal vision. These 'intrinsically photosensitive retinal ganglion cells', isolated by David Berson and Samer Hattar in 2002, form 2 per cent of the layer above the retina's rods and cones, and are connected directly to the master clock.

Clock genes

In one experiment, Bünning bred fruit flies in constant light for 30 generations (about one year) to destroy their circadian rhythms. When he put the flies in total darkness, the rhythms returned. This showed that organisms do not track time using some sort of memory; the clock is inherited genetically.

The gears of a body clock, its moving parts, are proteins whose levels oscillate over the course of the day. These proteins are encoded by 'clock genes'. The first gene was discovered in 1971 by American geneticists Ronald Konopka and Seymour Benzer, who identified three mutant forms of flies with abnormal patterns of movement and emergence: flies with circadian rhythms longer than the free-running period, those with a shorter period and flies with no rhythm. The mutations behind their behaviours mapped to the same location on DNA, a gene now called 'period'.

> Circadian rhythms reflect extensive programming of biological activity that meets and exploits the challenges and opportunities offered by the periodic nature of the environment.
> Colin Pittendrigh

Components of the body clock differ between species, but the basic mechanism is the same. As the activity of clock genes oscillate, levels of their encoded proteins rise and fall, which influences whether a protein attaches to DNA and switches a physiology-related gene on or off, which ultimately affects behaviour. For example, the clock gene 'tok1' controls when plants wake up in the morning and closes pores in leaves to prevent water loss in the evening. The clock components also interact and regulate each other in a feedback loop.

The condensed idea
Behaviour is synchronized to a daily rhythm

37 Sleep

The distinct resting period of sleep leaves an animal vulnerable to environmental threats, yet all species do it to some extent, with humans spending about one third of their lives asleep. Although there are several theories for sleep's function, it remains one of biology's biggest mysteries.

The features of sleep are easy to recognize: animals are physically inactive compared to their waking state, are less able to respond to most stimuli, and adopt a sleeping posture specific to their species. The state can be readily reversed, distinguishing it from dormant states like torpor or hibernation (lowered metabolism in the short- or long-term) and coma (the sometimes irreversible state of 'deep sleep'). Some form of sleep occurs in every species with a nervous system.

Mammals and birds have two distinct types of sleep: rapid eye movement (REM) and non-REM (NREM) sleep, which can be detected by electroencephalogram (EEG) recordings of brain activity. During NREM, nerve impulses appear as synchronized waves of electrical activity across the brain, whereas REM brain waves are chaotic and resemble waking activity. REM was discovered by American physiologists Nathaniel Kleitman and Eugene Aserinsky in 1953. A few years later, Kleitman and William Dement found that dreaming is associated with REM and that human sleep occurs in cycles that each consist of three stages of NREM followed by REM.

Animals are not normally active while sleeping because the brain paralyses the body, restricting movements to vital systems like breathing and twitches such as eye movements. The control system was revealed in 1959 by French neurobiologist Michel Jouvet, who observed cats with lesions in the pons region of the brainstem. When damaged, the pons is unable to inhibit motor centres in the medulla oblongata, which determine muscle atonia (paralysis). During REM, Jouvet's cats displayed behaviours such as attacking an invisible enemy – they were acting out their dreams.

Functions of sleep

As sleep researcher Allan Rechtschaffen once said, 'If sleep doesn't serve an absolutely vital function, it is the biggest mistake evolution

Sleep cycles

Over the three stages of NREM (non-rapid eye movement) sleep, brain waves become progressively slower and more synchronized and it becomes harder to wake an animal, with a trough at the N3 stage – slow-wave or deep sleep. Brain activity speeds up and enters REM sleep (sometimes with brief awakenings) before the cycle starts again. Each sleep cycle lasts about 90 minutes in adult humans.

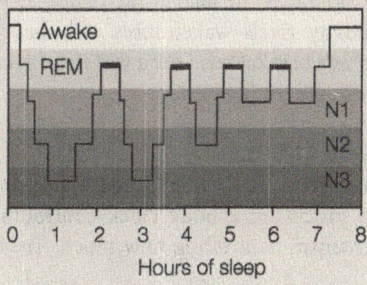

ever made.' His experiments in the 1980s showed that rats could be forced to stay awake (through fear of drowning) by putting them on a floating disc in a bucket of water but, after two or three weeks, the effects were lethal. In people, sleep deprivation impairs cognitive ability and changes mood and personality. And yet migrating birds can go for long periods without sleep, while pigeons subjected to the floating-disc experiment suffer no ill effects, suggesting that sleep is not always essential.

There are three main theories for why we sleep: conserving energy, repair and restoration, and maintaining the brain. The big brown bat, a nocturnal predator that sleeps 20 hours a day, seems to be saving energy rather than recovering from its four hours of hunting. It also makes intuitive sense that the body needs to fix itself and replenish the supply of molecules used when active. Finally, sleep probably reinforces and prunes the neural connections made for learning and memory – researchers such as Robert Stickgold have shown that recall is better after people sleep. The human brain constitutes 2 per cent of our total body weight yet consumes 20 per cent of calories used during quiet waking, so downtime could have multiple advantages.

However, every sleep theory has its weaknesses. If sleep conserves energy, why are animals tired after coming out of hibernation? If it repairs and restores, why does the body make more protein while awake? If it maintains brains, why do cetaceans (whales and dolphins, some of the most intelligent mammals) shut off just half their brain, for up to two hours? During this 'unihemispheric sleep', Jerome Siegel has found that cetaceans continue swimming without bumping into things, and have no REM. In fact, REM defies many explanations because brain activity rivals wakefulness. All these contradictions suggest that 'sleep' is a blanket term for a variety of processes that now coincide during a single resting period.

The sleep–wake cycle

In 1980, Swiss researcher Alexander Borbély proposed that sleep is regulated by two processes: a body clock controlling when, and a homeostatic mechanism controlling how much. The body clock is a

Dreaming

Being almost as mysterious as sleep, there are multiple theories for why we dream. One of the first came in 1977 from psychiatrists Allan Hobson and Robert McCarley, who suggested that dreams are simply a side-effect of random brain activity, which the mind tries to make sense of by stitching things together into a story. Another leading theory is that because dreams often feature recent events, they are involved in processing information. In mammals, studies by neuroscientist Matthew Wilson have shown that rats dream of waking activities (such as running through a maze) while asleep, as detected through neural activity in the hippocampus, the brain region that helps form memories. It remains unclear whether the narrative created in dreams is useful in memory consolidation, but as most dreaming occurs during REM sleep, it seems the reason we dream probably reflects the function of REM sleep itself.

timekeeper synchronized to external cues like light, while the homeostat turns sleep intensity up or down based on deficits or excessive sleep. Precisely how the homeostat works is not yet known, but deprivation prompts cells of the basal forebrain to release adenosine, suggesting that its level acts as a kind of 'counter'.

The master body clock and sleep homeostat are both found in the hypothalamus, an almond-sized region just above the brainstem that also controls appetite, thirst and other arousal functions. Specifically, both sleep regulators are located in a cluster of neurons called the SCN (suprachiasmatic nucleus).

As darkness falls, cells in the eye send signals to the SCN, telling the nearby pineal gland to release the hormone melatonin, which broadcasts a sleepy signal around the body. However, the real 'switch' of the sleep–wake cycle seems to be hormones known as hypocretins or orexins, which trigger the onset of wakefulness in the brain. These molecules, independently discovered in 1998 by two teams, are released by the hypothalamus and resemble secretin, a hormone that regulates water in the body.

> Sleep is a topic on which almost everyone considers himself an authority because of personal interest and firsthand experience.
> Nathaniel Kleitman

In the 1970s, William Dement established a dog colony at Stanford University to study sleep disorders. This led Emmanuel Mignot to study narcolepsy, a sudden onset of sleep, identifying a link to problems in detecting the wakefulness hormone, hypocretin. Dobermans carry a mutation in their hypocretin receptor, while narcoleptic humans have damaged hypocretin-making cells of the hypothalamus. Scientists are now aiming to create an artificial form of the hormone – not only for narcoleptics, but for anyone who needs to stay awake, such as pilots or long-distance drivers. We may not know why we sleep, but we may soon be able to control it.

The condensed idea
A resting state that plays several roles

38 Memory

We imagine that past experiences are recorded by a video camera and memories are movie files that can be replayed from a computer's hard drive. But a memory is not a continuous sequence of images, it consists of information scattered across different synapses – the gaps between brain cells.

Most animals have an 'implicit memory' for perception and motor skills, which includes reflex actions: in 1903, Russian physiologist Ivan Pavlov, an expert on digestion, noticed that dogs in his lab would not only salivate at the sight or smell of food, but also when his assistant appeared before feeding time. This so-called 'psychic secretion' inspired his famous experiment: ringing a bell around dinner conditioned the dogs to associate the two stimuli. They would then salivate from the sound alone, proving that a behavioural response could be modified through learning.

Animals with sophisticated nervous systems also have 'explicit memory', used for facts and events, which requires conscious awareness. In humans, both memory types – implicit and explicit – are stored in the cortex of the brain, but are formed using different regions – as illustrated by the case of 'HM', a 27-year-old American who suffered from seizures. In 1957, surgeon William Scoville and neuroscientist Brenda Milner described the effects of removing part of HM's medial temporal lobe, an area above the brainstem that includes a pair of curved structures – the hippocampus. After the procedure, HM could recall events from 19 months before, but not after. HM was cured of epilepsy but left with 'anterograde amnesia' – the inability to form new memories.

Learning

Until the mid-20th century, scientists tended to treat the brain as a black box. This is no surprise when you consider that a typical mammalian brain consists of billions of neurons, each with a thousand synapses – junctions connecting to other cells. Rather than try to untangle this complex wiring, Austrian-American neuroscientist Eric Kandel set out to study the simple: *Aplysia*, a sea slug with 20,000 neurons whose large, accessible cells made it easier to examine how learning modifies behaviour. Starting in the 1960s, Kandel used tiny electrodes to record impulses in a simple circuit of 30 neurons controlling a basic defensive reflex in the marine mollusc, where it withdraws the gill under its body when its siphon is touched.

Kandel began with a learning process called sensitization. Just as regular frights in a scary movie make you more sensitive to a harmless tap on the shoulder, Kandel found that mild shocks to *Aplysia*'s tail made it more sensitive to a touch on the siphon. The slug remembered the unpleasant experience, and that memory's duration depended on the frequency of shocks: it would forget within one hour after a single shock, but four single shocks caused it to remember for over a day. Converting a short-term memory to long term therefore involves spaced repetition.

Synaptic plasticity

In 1894, Spanish biologist Santiago Ramón y Cajal, the father of modern neuroscience, proposed that the connections at synapses were not fixed, but flexible, a principle now known as 'synaptic plasticity'. This was proven in Kandel's work: because he used the same neural circuit for different learning processes – including the opposite of sensitization, habituation – it was clear that memory is not encoded by the cells, but by their connections.

Explicit memory for facts and events is far more complex than implicit memory like reflexes, and sometimes involves connecting seemingly unconnected information. So how do the neurons become associated with one another? In 1949, Canadian psychologist Donald Hebb proposed that if one cell regularly fires an impulse towards a neighbouring cell, which then sends its own signal, the synaptic connection between them grows stronger. Simply put, 'cells that fire together, wire together'. The physical effect is now called long-term potentiation (LTP).

An individual memory is like a river that only becomes visible when water is flowing, when cells are firing. Its physical trace (the 'engram') is an imprint left in a dry riverbed, which can be made deeper through long-term firing of neurons. The riverbed itself is made of molecules: proteins inside neurons, calcium ions creating voltage across cell membranes, and neurotransmitters such as serotonin, glutamate and dopamine in the synaptic gaps.

Memory is unreliable, malleable and easily modified. In 1974, American psychologist Elizabeth Loftus found that false memories can be implanted through a 'misinformation effect': after asking a leading question or changing a tiny detail, like substituting a neutral word like 'hit' with 'smashed', witnesses of a car crash had distorted recall and remembered seeing broken glass at a scene where none existed. This suggests that people should not be convicted of a crime based on eyewitness testimony alone.

Scientists are now aiming to exploit the unreliable nature of memory to edit bad experiences and improve mental health, such as reducing the power of flashbacks that soldiers get through post-traumatic stress disorder (PTSD). Unlike computer files organized in a folder, a memory cannot be deleted in one click because its bits – including sensory information and associated emotions – are scattered around the brain. But it can take a while for an event to be fully consolidated into long-term memory, enabling people to 'remember to forget': asking a PTSD patient to relive a recent trauma while giving them a drug like propranolol (a beta blocker) interferes with the molecules that form and maintain memories. This can cut synaptic connections between an event and its associated stress, reducing the emotional impact of the memory.

Storage and recall

Short-term memory retains information for seconds or minutes; long-term memory may last for years. In 1986, Kandel revealed the molecular difference between them: after giving *Aplysia* cells various drugs that block the production of new proteins, short-term sensitization remained but long-term memory did not. Short-term memory therefore uses pre-existing molecules at the synapse whereas long-term requires new protein synthesis, which involves communication between synapses and the cell nucleus along a molecular chain reaction and the gene-activating proteins CREB-1 and CREB-2.

Much less is understood about how memories are maintained and recalled. One interesting finding is that recent experiences go through a period of 're-consolidation' before becoming stable, long-term memory. In 2000, Canadian neuroscientist Karim Nader taught rats to associate a mild shock with a high-pitched tone. After the animals were injected with a drug that prevents new memories from being formed, they failed to flinch upon hearing the noise – the rats forgot their fear. The word 'recollection' best describes how a memory is retrieved: bits are accessed and recombined each time. Far from being recordings, memories are easily manipulated.

> If elementary forms of learning are common to all animals with an evolved nervous system, there must be conserved features in the mechanisms of learning at the cell and molecular level.
> Eric Kandel

The condensed idea
Experiences are stored as a pattern of neural connections

39 Intelligence

ntelligence is the product of cognition, the mental processes that allow an animal to acquire and apply knowledge. By that definition, humans are the smartest species, so one way to identify other clever creatures is to compare their cognitive abilities to our own, overturning some animal stereotypes.

Animal intelligence cannot be measured by IQ tests because other creatures cannot read or write – and the tests require human biology, such as having opposable thumbs. This anthropocentric bias is important to bear in mind when comparing species.

It was once assumed that our species' tool-making ability was a defining and uniquely human trait. But then British primatologist Jane Goodall started studying chimpanzees in Tanzania. One day she noticed a male chimp stripping a stick of leaves and using it to extract termites from mounds. After Goodall reported this observation in 1964 to her mentor, anthropologist Louis Leakey, he replied: 'Now we must redefine man, redefine tool, or accept chimpanzees as human.'

Making tools

Today, we know that tool use occurs across the animal kingdom. Bottlenose dolphins in Australia's Shark Bay protect their beaks with sponges while foraging on the sandy ocean floor, for example, while capuchin monkeys pick stones to dig and crack nuts. A key aspect is that animals must alter and hold the device – a branch only becomes a tool when detached from a tree. Many tools are found objects, and a more impressive feat is to modify something to improve its function – what you could call technology.

Only a few species manufacture sophisticated tools. The skill was limited to primates until comparative psychologist Gavin Hunt visited the islands of New Caledonia in the South Pacific. Hunt observed that one crow species would bend twigs into hooks and use leaves with tiny barbs along their edge, tearing strips in deliberate steps to fashion long, tapered tools that can snag grubs in tree holes. The 'bird brain' stereotype has now been overturned by New Caledonian crows and other members of the corvid family.

Understanding language

Parrots repeat words, but do they understand them? After a PhD in theoretical chemistry, Irene Pepperberg decided to find out. In 1977, she bought a one-year-old African grey parrot from a pet shop near Chicago airport and named him 'Alex' (supposedly standing for 'Avian Learning Experiment'). Over three decades, Pepperberg taught the bird over 100 words. Shown a green key and green cup and asked what was different, Alex said 'shape'. Asked what was the same, he would reply 'colour'. He could count to six and improvised when he struggled. Because a red apple tasted like banana and looked like a cherry, he called it 'banerry'. Alex was capable of more than simply 'parroting'.

Talking animals are rare because most species do not have lips, vocal cords or other features to mimic human speech. Several non-human apes have, however, been taught to communicate by other means. At San Francisco Zoo, Koko the gorilla knew over 1,000 sign-language gestures, while American primatologist Sue Savage-

Consciousness

'What is it like to be a bat?' This question was asked in 1974 essay by philosopher Thomas Nagel, who argued that life as a flying mammal, using echolocation to navigate, is so alien to human experience that we can never understand how a bat perceives the world. Given that philosophers have been debating consciousness for centuries, can biology help solve it? Nagel's essay was an argument against 'reductionism', the view that complex systems like the brain can be explained as the sum of their parts. But many scientists believe this is a practical way to tackle the 'hard problem' of consciousness, how subjective phenomena have certain qualities like colour or taste ('qualia'). Because experiences are ultimately encoded by the behaviour of nerve cells, it should in principle be possible to detect associated events or patterns in the brain – the 'neural correlates of consciousness'. Neurobiologists now believe that consciousness can be solved, starting with what it is like to be a human.

Rumbaugh taught Kanzi the bonobo to recognize lexigram symbols on a board or touchscreen. Dolphins can take things to the next level by comprehending the specific order of words in a sentence – its syntax. As reported by Louis Herman in 1984, Atlantic bottlenose dolphins can use simple grammar with gestures: a female called Akeakamai knew that a closed-fist pumping motion was 'hoop', arms vertically overhead was 'ball' and a come-here gesture meant 'fetch'. If told 'hoop-ball-fetch', Akeakamai would push a ball through a hoop, whereas 'ball-hoop-fetch' prompted her to bring hoop to ball. She could also tell right from left.

> I knew it was very exciting ... to pick a leafy twig and strip the leaves off, which is the beginning of tool making.
>
> Jane Goodall

Self and others

In 1970, psychologist Gordon Gallup let chimpanzees get used to a mirror before anaesthetizing them and marking their faces with red dots. When the chimps woke up and saw their reflection, they reached for the marks, recognizing their own appearance, whereas monkeys reacted as if their reflection was a new individual. Dog owners may claim their pets know what they are thinking, but the animals fail this 'mirror test'. This does not necessarily mean they are not self-aware, however: a dog's main sense is smell, not sight. Nonetheless, animals like dolphins, elephants and the European magpie (a corvid) do pass the test. Apes – including human babies – become self-aware at 1–2 years old.

What is another animal thinking? In psychology, this is 'theory of mind' – the ability to understand that another individual's mental state can differ from your own. In 2001, British cognitive scientists Nathan Emery and Nicola Clayton showed that Florida scrub jays can remember specific events – episodic memory or 'mental time travel'. If a bird sees that a competitor has watched it hide food, it moves its cache while the second bird is away. This is common when a jay has previously been a thief itself, suggesting it understands intent to steal.

Bigger and better brains

Why are some species smarter than others? The most intelligent have larger brains relative to their body size, which matches anecdotal evidence from tool use, language and self-awareness: dolphins and primates are smart compared to sheep and mice; parrots and corvids are clever, pigeons and chickens are not. But intelligence depends not only on brain-to-body mass ratio, but also on neural structures.

The grey matter of a mammal's brain resembles a walnut because of its folded neocortex, while bird's grey matter is organized into pockets. The structures are different but equivalent, which feeds a need for brain power. As Nathan Emery and Nicola Clayton describe it, 'The analogy would be to compare a club sandwich (mammalian) to a pepperoni pizza (avian).' The pair propose that corvids and apes have comparable abilities due to similar social environments – to deceive competitors, for example. Ecological pressures then drove natural selection, leading to convergent evolution of equivalent brain structures in distantly related species. This is why crows can be as clever as chimpanzees.

The condensed idea
Ecological pressures shape cognitive ability

40 Humans

What makes us human? Where are we from? Why are we different? Until about a decade ago, such questions could only be answered by archaeologists and anthropologists, but sequencing technology now enables biologists to compare the DNA of *Homo sapiens* with other species – including our extinct ancestors – to reveal the secrets of human origins.

Each species is special, with unique adaptations that set it apart from others, yet there is no doubt humans have features that are unrivalled compared to the rest of life on Earth, such as the ability to transmit knowledge through language. According to archaeologists, our ancestors started making stone tools 2.6 million years ago, while 'anatomically modern humans' came out of Africa some 200,000 years ago. Palaeo-anthropologists say that cultural artefacts like art appeared across Eurasia 60,000 years ago, when humans started to colonize the globe.

Great apes

Humans are apes, but our species is clearly distinct from other members of the great ape family. One example is spoken language, an ability that highlights the challenge of finding the genetic factors underlying what makes us different. In 2002, scientists reported that a gene that causes speech and language disorders if mutated, FOXP2, seemed to have evolved relatively quickly in humans compared to chimpanzees. This implied that natural selection had favoured changes to its DNA, helping to explain our vocal communication skills. But in 2018, researchers revealed that Neanderthals (a species not believed to be great talkers) had the same changes to FOXP2.

The full chimpanzee DNA sequence was read in 2005 with the aim of finding differences from our genome. Gorillas and orangutans soon followed, with macaques (Old World monkeys) for contrast. Although aligning genomes side-by-side has revealed some differences, identifying those important to human evolution has been hard. Comparing our genome to the chimpanzee shows that since we shared an ancestor, our DNA has accumulated 20 million nucleotide substitutions (single-letter changes). It sounds like a lot, but it only represents 0.6 per cent of our genome's 3.2 billion base pairs.

One way to identify potentially important DNA is to detect sections that have been gained or lost. In 2011, a team led by Gill Bejerano and David Kingsley found over 500 deletions in humans relative to other primates, and studied two in detail. One was the loss of an enhancer (DNA control element) for an androgen receptor gene, which had the lovely anatomical effect of removing tiny spines from the penis. An enhancer was also lost from GADD45G, a gene that restricts cell division in the cerebral cortex, suggesting that this deletion might have allowed the human brain to increase in size.

Archaic differences

All human-like apes since the split from chimpanzees – including species of *Homo* and *Australopithecus* – are 'hominins', while the extinct hominins of the past half-million years are called 'archaic' humans. Neanderthals are the most famous example, and first appear in the fossil record over 300,000 years ago, going extinct around 30,000 years ago. Everyone knows that their body was beefier than modern humans, but their brains were bigger, too. What gave *Homo sapiens* an advantage over archaic humans?

Progress in so-called palaeogenetics has long been plagued by technical problems such as contamination, but in 2010, a team led by Swedish geneticist Svante Pääbo finally published the Neanderthal genome. The most controversial finding is that around 2 per cent of our own genome is closely related to Neanderthal, suggesting that we interbred. In 2010, the genome from a small fragment of finger bone found in the Denisova Cave, Siberia, revealed another unknown species, the first group of archaic humans to be defined without a skeleton. These 'Denisovans' contributed about 5 per cent of genetic material to modern people from the south Pacfic.

> Sequencing of ancient genomes will tell us why, of all primates, modern humans were the ones to spread to all corners of the globe and reshape the planet.
>
> Svante Pääbo

According to Pääbo, modern humans did not fully replace the archaic species – initial interbreeding meant a 'leaky replacement'. Almost two billion letters from the human genome can be aligned to Neanderthal and Denisovan DNA, revealing several surprises about human evolution. For example, the changes in FOXP2, linked to brain

People are incredibly similar at the DNA level: 99.9 per cent identical for single-letter differences. But those numbers tell us little about how genetic variants – alleles – make us unique. How could biologists offer insight into genetic individuality? One way is to introduce human alleles into mice, an example being the EDAR (ectodysplasin receptor) gene, where the 370A variant produces thicker hair and shovel-shaped teeth. This variant is carried by close to 100 per cent of people in many Asian populations, and arose in China 30,000 years ago. In 2013, a team led by Bruce Morgan and Pardis Sabeti created genetically engineered mice with 370A. As well as thicker fur, the animals also experienced other changes, including extra sweat glands. Surveying Han Chinese people, Morgan and Sabeti found that 370A is associated with making more glands in humans too. Genetic modification is also used to reveal what makes our species unique. In 2021, Alysson Muotri grew organoids (organ-like structures) from cells of the human frontal cortex, in which a modern version of the NOVA1 gene was edited to have archaic variants found in Neanderthals and Denisovans. These 'mini-brains' developed differently, with altered neural networks (formed by synaptic connections between cells) – hinting at why we became smarter than our extinct relatives.

function, predate the split from archaic humans. We differ from archaic humans by 32,000 single-letter changes, and a 2013 study led by Pääbo and David Reich linked some of the proteins involved to the early development of the cerebral cortex.

Modern variation

As humans spread across the globe, populations adapted to local environments and created present-day diversity. So we might expect people from the same continent to be genetically similar, but this is not the case. In 2010, the 1000 Genomes Project compared the genomes of 185 individuals from two African populations with 184 from Europe and China. Although this revealed almost 39 million positions where the DNA varied, not a single one was common to all Africans or all Eurasians.

Genetic difference in populations with broadly similar characteristics can happen because such features are not down to a specific genetic variant, but to interaction *between* variants. Take height, for instance. In Mendel's pea plants (see chapter 7), height is determined by two variants in one gene, but human height is influenced by at least 180 positions in DNA. In 2012, the Genetic Investigation of Anthropometric Traits (GIANT) project found that 85 of 139 height-increasing variants are more common among northern Europeans than in southern populations.

Adaptations result from natural selection. For many features, like increased height, we may never know whether they were shaped by the environment or mate choice. Other features are obvious. For example, one variant of the G6PD gene is found in one-fifth of people where malaria is prevalent, as it confers 50 per cent immunity to parasite infection. Then there is skin colour: a variant of the SLC24A5 gene is linked with lighter pigmentation and is common in Europe. How do all these variants combine to create a unique human being? That is a question geneticists hope to answer over the next decade.

The condensed idea
The secrets of our species lie in the genome

41 Pollination

Flowering plants – angiosperms – dominate Earth's terrestrial vegetation, creating every habitat from temperate grassland to tropical rainforest. The majority of angiosperm species exploit animals to disperse their pollen, a reproductive strategy that has become vital to plants and crucial to human agriculture.

Flowering plants provide most of mankind's nutrition. The crops that constitute half our diet – cereals such as rice, maize and wheat – disperse pollen via water or wind (abiotic pollination), but three-quarters of crop species – those supplying most of our fruit and vegetables – benefit from being pollinated by bees and butterflies, bats and birds and many other animals. This close relationship is incredibly popular: a 2011 survey by ecologist Jeff Ollerton found that almost 88 per cent of flowering plants – over 300,000 angiosperm species – reproduce via pollinators.

Flower theory

When you think about it, biotic pollination is quite kinky: one kingdom of life uses organisms from a different kingdom for a helping hand during sex. Even the idea that plants have sex was once considered scandalous, which followed observations from a series of German botanists, starting in 1694 with Rudolf Jakob Camerarius, who described the male and female reproductive parts of the flower. In the 1760s, Joseph Gottlieb Kölreuter described pollen and transferred grains between plants to create hybrids, suggesting a potential role of insects in cross-fertilization.

Christian Konrad Sprengel turned pollination biology into a science with his 1793 book *The Secret of Nature Discovered in the Structure and Fertilization of Flowers*. Studying over 460 species, he developed the idea that floral features seem designed to attract insects. Before Sprengel, most botanists believed animals visited flowers by accident, so things like nectar were somehow useful to plants. Among many proposals, Sprengel stated that nectar guides – colour patterns on petals – direct insects towards the sweet reward, via a brush with sticky pollen grains. Sprengel also saw that flowers deceive insects, so plants are puppet masters and animals are their marionettes. Sprengel's work became widely known through Darwin, whose 1862 book, *Fertilization of Orchids*, focuses on a family that makes up a tenth of angiosperms. Scientists have since revealed that flowers supply pollinators with a variety of nutrients – along with carbohydrates from nectar, pollen is a source of protein, for example. We now know that some relationships are exclusive, as in yucca plants and 'yucca moths', but many plants are far more promiscuous, using an array of animals to disperse pollen.

Plant life

While animal sex involves direct encounters between parents or gametes (sperm or egg), reproduction in land plants is a more complicated affair. Their life cycle consists of alternating generations: a gametophyte that makes gametes carrying a single, haploid set of genes; and a sporophyte generation that produces spores with a diploid genome. Ancient plants such as ferns and mosses spread all spores, but angiosperms only disperse pollen – the male microspore – while female megaspores – ovules – grow from gametophytes that are housed and fed by their sporophyte parent. Fertilization occurs after pollen lands on the female part of a sporophyte, forming a seed.

> That [bees] and other insects, while pursuing their food in the flowers, at the same time fertilize them ... appears to me to be one of the most admirable arrangements of nature.
>
> Christian Konrad Sprengel

Seed plants are either angiosperm ('seed vessel') in Greek, or gymnosperm ('naked seed'). In angiosperms, the parts around an ovule become fruit, a delicious vessel that encourages animals to disperse seeds by throwing away the large core or allowing small pips to pass through the gut. Gymnosperms include softwood trees whose seeds are protected by cones (conifers and cycads), plus the 'living fossils' like gnetophytes and a group containing one species, *Ginkgo biloba*. A 2009 review by evolutionary botanists William Crepet and Karl Niklas showed that gymnosperms make up 0.3 per cent of living plant species, while flowering plants constitute almost 90 per cent. A gymnosperm life cycle is very slow, with pollination to fertilization taking a year or more, and generation time (from seed to seed) lasting centuries. By contrast, the first plant to have its complete genome sequenced – *Arabidopsis thaliana*, a white-flowered member of the mustard family – has a life cycle of 1–2 months. This helps explain why angiosperms dominate the plant kingdom.

Angiosperm diversity

For plant lovers, the most exciting event in life's history is not the Cambrian explosion of animals, nor the extinction of the dinosaurs, but the 'angiosperm radiation' – the rise in flowering plant diversity during the Cretaceous period. After appearing in the fossil record 130 million years ago (MYA), angiosperms became widespread and diverse by 100 MYA. Their spread seems so rapid that, in 1879, Darwin called it an 'abominable mystery'. Biologist William Friedman argues that Darwin saw the rapid rate of angiosperm evolution as a specific case that posed a general problem for his theory, because he believed that change could only be gradual. If there were sudden jumps – saltations – then they could be interpreted as creation. In an 1881 letter, Darwin clarified his view on the rise of angiosperms by saying it was *'apparently* very sudden or abrupt' because fossil evidence is never complete.

Crepet and Niklas's comparisons find no difference in rates of speciation, extinction or diversification among angiosperms, gymnosperms and ferns over the last 400 million years – so why did angiosperms become so diverse? In an 1873 book and later correspondence with Darwin, French palaeontologist Gaston de Saporta proposed a link with the evolution of pollinating insects and flower arrangements. This was supported by Crepet and Niklas, who found strong correlations between the number of angiosperm species, floral features and insect families. This does not mean that plant radiation drove insect diversity (or vice versa), but supports Saporta's idea of coevolution. One potential cause is that plants can double their genome with few ill effects, allowing duplicate genes to evolve new functions. While Crepet and Niklas found nothing unusual about the rate of species changes, sustained speciation has allowed flowering plants to continuously reinvent themselves.

Colony collapse disorder

In 2006, American beekeepers started reporting the mysterious disappearance of their insects: queen bees stayed in the hive, but most workers vanished. Colony Collapse Disorder (CCD) has been blamed on everything from parasitic mites to habitat loss, but the main culprit is a class of nicotine-like pesticides – the neonicotinoids or 'neonics' – which are sprayed on crops and end up in plant cells. In 2012, Dave Goulson found they cause bumblebee colonies to grow slower and make fewer queens, while Mickaël Henry tracked honeybees using RFID tags and found that neonics (neurotoxins) interfere with homing ability. Then, in 2015, Clint Perry altered the age structure of colonies and replicated CCD symptoms without chemicals, which seems to solve the mystery of CCD. Older bees normally handle foraging duties while the young carry out housekeeping tasks, but when the older insects fail to return, others have to pick up the slack. Because the youngsters are not good at finding food, the whole colony suffers from the stress of starvation. If bees do not learn to forage before the backup supplies run out, a colony collapses. CCD is therefore caused by the loss of experienced bees, triggered by neonics.

The condensed idea
A mutually beneficial relationship for insects and plants

42 The Red Queen

cological interactions can be positive, as in pollination, but many are negative – antagonistic relationships between predators and prey, parasites and hosts. The Red Queen hypothesis is one of the most influential concepts in biology and helps explains why conflict drives the coevolution between two species.

In *Through the Looking Glass*, Lewis Carroll's sequel to *Alice's Adventures in Wonderland*, Alice races to catch the Red Queen only to discover that neither of them has moved. The Queen explains that in her country 'it takes all the running you can do, to keep in the same place'. In recent times, this has been used as a metaphor for why natural selection drives antagonistic coevolution: a species must ceaselessly adapt in response to its adversary's adaptations.

Constant extinction

The Red Queen hypothesis was proposed in 1973 by American evolutionary biologist Leigh Van Valen, an eccentric polymath who wrote songs with titles like 'Mexican Jumping Genes' and 'Sex Among the Dinosaurs'. After studying various fossils, Van Valen found the rate of extinction was constant regardless of geological lifespan. After his paper 'A New Evolutionary Law' was rejected by academic journals, he published it by launching his own journal, *Evolutionary Theory*.

Van Valen used the Red Queen hypothesis to explain his 'law of constant extinction' – species must keep adapting regardless of their age – and suggested that conflict between species creates an ever-changing environment that drives evolution by natural selection. Van Valen called this a zero-sum game: there are no winners, only losers that go extinct. His metaphor has since been used to explain various phenomena, most famously sex, as argued by evolutionary biologists John Jaenike and W.D. Hamilton. The original concept involved members of two species, but the Red Queen can also apply to the conflict between parents and offspring, battle of the sexes, and selfish genetic elements.

The Red Queen creates natural enemies. Conflicts are ultimately a fight over an ecosystem's resources, especially food, leading to antagonistic interactions between an 'exploiter' that steals resources from a 'victim'. These exploiter-victim relationships include every host-parasite, predator-prey and plant-herbivore interaction. Direct conflict between plants and herbivores are not clear, however, because there are more than two antagonists: veggies are eaten by multiple species. Meanwhile, parasites are often adapted to a single host. The links between a parasite's weapons and a host's defences – seen in their physical features and genetic variants – is visible as an arms race between them.

Evolutionary arms races

Host–parasite relationships are clear cases of the Red Queen in action, as shown by humans and *Mycobacterium tuberculosis*, the pathogen that causes TB. In 2014, microbiologists sequenced 259 genomes to reconstruct the bacterium's evolutionary history and found it emerged 70,000 years ago, after humans migrated out of Africa. It became genetically diverse as population density increased during the late Stone Age. In 2005, a comparison against chimpanzee DNA showed that the gene for granulysin – an antibiotic that attacks TB – is evolving rapidly in humans, which suggests an arms race. The weapons we steal from other species, like penicillin from mould (discovered by Alexander Fleming in 1928) help fight parasites, but our foes also run with the Red Queen, creating antimicrobial resistance to drugs and superbugs like MRSA.

> Each species is part of a zero-sum game against other species. Furthermore, no species can ever win, and new adversaries grinningly replace the losers.
> Leigh Van Valen

Prey–predator relationships are arms races, but the conflict is often obscured because natural selection applies unequal forces on each competitor. As explained by British biologists Richard Dawkins and John Krebs in 1979, 'The rabbit runs faster than the fox, because the rabbit is running for his life while the fox is only running for his dinner.' This 'life-dinner principle' shows the penalty of failure, and why mutations that cause rabbits to lose are unlikely to spread through a gene pool: 'No rabbit has ever reproduced after losing a race against a fox. Foxes who often fail to catch prey eventually starve to death, but they may get some reproduction in first.'

Selection influences the evolution of natural enemies in three main ways. First, an arms race can escalate, sometimes leading to exaggerated weapons and defences, such as the long snout of weevil beetles and the thick fruits in *Camellia* plants. This 'escalatory Red Queen' coevolution leaves its mark in the fossil record, what Dutch palaeontologist Geerat Vermeij calls the escalation hypothesis. The second arms-race scenario, 'chase Red Queen', occurs when a victim species is under strong natural selection pressures to evolve novel features that force an exploiter to keep up. Third, 'fluctuating Red Queen' effects happen when the frequency of gene combinations – in exploiters and victims alike – repeatedly wax and wane over time.

An end to war

After telling Alice 'it takes all the running you can do, to keep in the same place', the Red Queen adds: 'If you want to get somewhere else, you must run at least twice as fast as that!' So how do organisms escape conflict? Prey migration might force a predator to find fresh meat, for example, or hosts might develop total immunity against a parasite. But if exploiters kill too many victims, it can lead to mutual extinction, so levels of virulence or predation play a part in the outcome of conflicts. Fights can also be temporary truces rather than permanent peace, like between ourselves and some bacteria in human microbiota.

The Red Queen can explain conflict between two antagonists, but the multiple interactions within a community or ecosystem are far more complex. In 1999, palaeobiologist Anthony Barnosky suggested that extinction and speciation rarely happen except in response to environmental change. Playing on the royal theme and nature's unpredictability, he named this the Court Jester hypothesis. The two hypotheses are not mutually exclusive, however, as natural selection is caused by both biotic and abiotic forces.

Human microbiota

In 2012, scientists from the Human Microbiome Project revealed the true biodiversity living in an intimate relationship with us. Researchers using DNA sequencing showed that microbiota colonize the human ecosystem to exploit our energy resources, and especially the carbohydrates we produce. Thousands of species occur in the gut, for instance, and bacterial cells outnumber human cells by an order of magnitude. The ecological interactions between us and them will vary by species, but most microbes are probably 'commensalists' that benefit from our resources without causing harm. Some will be parasites that affect health, others 'mutualists' – we give them a home; they protect us from pathogenic invaders that cause disease. Note that although we label our microbiota with names like 'friendly bacteria', some are potential enemies. From the exploiter-victim interactions of the Red Queen, exploiters can be 'obligate' parasites that harm as a consequence of their life cycle, or 'facultative' parasites that take advantage when an opportunity presents itself, like when immunity is compromised.

The condensed idea
Coevolution driven by conflict between natural enemies

43 Ecosystems

From lakes and deserts to rainforests and reefs, every habitat includes a web of interactions that allows energy and biomass to flow through the environment. These ecosystems contain limited resources, creating competition both within populations and between communities – a major driving force for natural selection.

Near the end of *On the Origin of Species*, Charles Darwin describes life as an 'entangled bank' where organisms are 'dependent on each other in so complex a manner'. The zoologist Charles Elton expanded on the idea of complex interactions in his 1927 book *Animal Ecology*, saying that each species has its own 'niche' – 'its place in the biotic environment, its relations to food and enemies'. But as fellow British ecologist Arthur Tansley pointed out in 1935, the environment includes inorganic parts, too. He proposed that biological and physical factors interact within ecosystems: 'basic units of nature on the face of the Earth'.

Ecosystems are battlegrounds in a war over niches, where life is ultimately fighting for energy. Earth's main energy source is sunlight, converted to biomass by photosynthesis in green algae and land plants. These 'producers' trap energy in the bonds of carbohydrate molecules, while 'consumers' release energy from carbohydrates through respiration, returning carbon and other elements to the biosphere. Energy is transferred from producers to consumers (and from primary to secondary consumers) by organisms eating one another.

Food webs

The top of the food chain is actually the tip of a pyramid. In 1942, American ecologist Raymond Lindeman grouped all species at the same positions in a food chain into levels of 'trophic pyramids' ('nourishment' in Greek). Autotrophs make their own food at the base, heterotrophs eat others, and saprotrophs like soil bacteria and fungi decompose organic matter in the pyramid's unseen foundations. Energy is lost via heat and waste during transfer, so the average trophic efficiency is only 10 per cent. This explains why ecosystems contain many plants but few apex predators, and why food chains are short, usually with 4–5 species.

Energy and biomass are transferred through a food web or pyramid. The top two nourishment or 'tropic' levels are consumers (such as carnivores and herbivores) and the base consists of producers (the foundation of decomposers is not shown). Food webs include multiple food chains, with nodes for 'species' and links for who-eats-whom. Strong interactions (heavy lines) could represent an exclusive predator prey relationship.

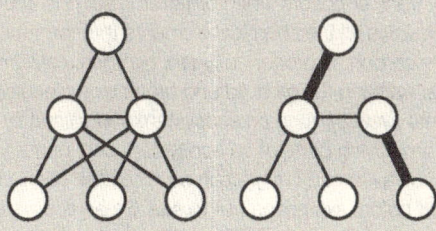

Food chains were first linked together into a web by Charles Elton in 1927. Food webs represent Darwin's tangled bank, and are now built with mathematical models to help answer questions about an ecosystem's complex interactions. For example, how do invasive species cause extinctions? What are the impacts of habitat destruction and anthropogenic climate change?

Diversity and stability

Why should we save species? Environmentalists assume 'more is better': based on observations from nature, Elton claimed that simple communities are more easily upset than rich ones. One example is cultivated land, where humans reduce biodiversity, making it more vulnerable to invasive species. In 1955, ecologist Robert MacArthur argued that populations are less susceptible to drops in either predators or prey if there are multiple predator-prey relationships.

But in 1973, Australian theoretical ecologist Robert May challenged these intuitive arguments. He built food webs using mathematical models in which the strength of interactions between species was randomly

Energy flow

Unlike habitats, ecosystems are not geographic locations. The ecosystem concept is an analogy: organisms are the moving parts in a machine powered by energy. So ecosystems are thermodynamic systems governed by the laws of physics. As Einstein's equation $E=mc^2$ shows that energy and mass are equivalent, while the law of conservation of mass states that 'matter cannot be created or destroyed', ecologists can study the flow of energy or biomass as though they were the same thing. Energy transfer occurs when organisms eat one another, which also provides all the chemical elements that form the body – mainly carbon, hydrogen, oxygen, nitrogen, calcium and phosphorous. As living things build and break biomolecules, those elements cycle through an ecosystem and ultimately the biosphere. The moving parts of an ecosystem continuously cross boundaries, however: birds migrate in summer, fish grow up on a coral reef before swimming out to sea. So an ecosystem is not a true, closed thermodynamic system, but is treated like one to study its complex interactions.

assigned (a strong link might represent predators that only eat one kind of prey). May's systems were *less* stable when they contained more connections, implying that stability is determined by specific interactions in nature.

Field ecologists took a different tack. In 1982, David Tilman began an 11-year study of stability at a single trophic level, the biomass produced by plants in an area of Minnesota grassland. His results showed diversity helping to maintain the food pyramid, at least at its base. Field tests at multiple trophic levels are fiendishly complex, but small-scale studies on microcosms of bacteria and protists also suggest diversity provides stability.

There is just one problem. Well, two: ecologists do not agree on definitions of either diversity or stability. Different species are often combined into one 'functional group' to simplify the web of interactions, and stability has multiple meanings. An ecosystem can have 'resistance' to change despite shifting environmental conditions, and can have 'resilience' to fluctuations, returning to normal after being disturbed. A stable system is not static either. Some lakes flip between two states – crystal clear or covered in scum – reflecting a battle between different algae. The idea of a 'balance of nature' is simply not scientific.

> Though the organisms may claim our primary interest ... we cannot separate them from their special environment, with which they form one physical system.
>
> Arthur Tansley

Interactions and insurance

Elton's argument for complex food webs and May's models showing that complexity is unstable can be reconciled by the way species interact. In 1992, Canadian ecologist Peter Yodzis compiled data from real food-web relationships to build models with plausible interactions, which revealed that the strength of interactions is key to stability. Strong interactions such as a predator feeding exclusively on one kind of prey could lead to runaway consumption, so stable ecosystems need numerous weak interactions, such as omnivores.

What is more, while some organisms are essential to an ecosystem, others may not be. In 1999, theoretical ecologists Shigeo Yachi and Michel Loreau outlined an 'insurance hypothesis': greater diversity raises the odds that at least some species will respond to environmental change, and increases the chance that a functional group contains a species capable of replacing an important species (so-called redundancy). Nevertheless, it is hard to predict which ones are essential to an ecosystem, and which ones are more easily replaced, so the safest approach is to assume each species is sacred. Ethical ideas like a moral obligation to preserve species might not convince governments, but the best argument for protecting ecosystems is practical: they are our life-support systems too.

The condensed idea
Stable food webs have weak interactions and diverse species

44 **Natural selection**

The theory of evolution by natural selection explains how everything from birds to bacteria adapt to their environment, and ultimately helps explain biological diversity. Today, it is often associated with one man, Charles Darwin, but it could easily have been credited to Alfred Russel Wallace.

A year before publishing *On the Origin of Species*, Darwin received a package in the post, containing an essay by a young naturalist, Alfred Russel Wallace, and a letter asking for his thoughts. It was 18 June 1858 and Darwin was at home in Kent, gathering evidence to support his theory that a struggle for survival leads to evolution. He opened Wallace's package and read the essay: it outlined almost exactly the same theory.

Darwin was devastated. He had recently told botanist Joseph Hooker there was no rush to read the manuscript for his 'big book' on species; now he was writing to geologist Charles Lyell, distraught. Wallace was in Southeast Asia but Darwin refused to treat him unfairly, saying: 'I would far rather burn my whole book'. Darwin had other things to worry about (his son was sick with scarlet fever) so Hooker and Lyell hatched a plan. On 1 July 1858, they presented two papers at the Linnean Society in London: Wallace's essay and an extract from Darwin's book. After Hooker and Lyell revealed their actions, both Darwin and Wallace said they were happy.

Origin of a theory

How did Darwin and Wallace come to propose the same theory? One common inspiration was biodiversity: Darwin spent five years (1831–1836) on his round-the-world voyage aboard HMS *Beagle*, studying geology and nature, Wallace earned a living collecting specimens and spent four years in the Amazon then eight years (1854–1862) exploring islands of the Malay Archipelago. Both read the Reverend Thomas Malthus's 1798 *An Essay on the Principle of Population*: this suggested that when population growth is faster than food production, numbers are kept in check by factors like famine and disease. This inspired the idea of competition for limited environmental resources.

For Wallace, natural selection was a 'Eureka!' moment that struck during a fit of malarial fever in Indonesia. For Darwin, it was a slow realization, as illustrated by his thoughts on birds he collected from the Galápagos Islands off the west coast of South America. Now known as Darwin's finches, the birds are dark-coloured with slightly different beaks. They barely get a mention in the 1839 book *The Voyage of the Beagle*, but by the 1845 edition, Darwin says: 'Seeing this gradation and diversity of structure in one small, intimately related group of birds, one might really fancy that from an original paucity of birds in this archipelago, one species had been taken and modified for different ends.'

Evolution in action

Natural selection is not always a slow and gradual process. It can be observed in a human lifetime, one example being Darwin's finches. Since 1973, Peter and Rosemary Grant have tracked birds on Daphne Major, a tiny island in the Galápagos where around 150 breeding pairs are challenged by environmental pressures from the El Nino-Southern Oscillation, a climate phenomenon that periodically flips atmospheric pressure and temperature. After a drought caused small seeds to become scarce in 1977, only birds with large beaks could crack open the nuts. Less than 20 per cent of the medium ground finch species survived, but in 1978 the average beak size among the offspring was 4 per cent higher. Natural selection in one year.

Another example of evolution in action is Richard Lenski's long-term evolution experiment. Since 1998, his lab has been growing 12 populations of *E.coli* in culture. Every 500 generations (75 days), some bacteria are transferred to new flasks while others are frozen as a record for that point in time. In 2008, Lenski and Zachary Blount discovered that one population had evolved the ability to eat citrate – a molecule the microbes cannot normally use as an energy source. This was later revealed to be the result of several random mutations. By around 32,000 generations (four years), the citrate-eating population could grow larger and had more genetic diversity.

The first coherent theory of evolution was proposed in 1809 by Jean-Baptiste Lamarck, who claimed that 'transmutation' of species occurred after organisms acquired characteristics during their lifetime. Darwin proposed that species changed via the survival of individuals that were already adapted to their environment, based on his observations of exotic island diversity and of domesticated species like dogs, horses and pigeons – the phrase 'natural selection' relates to selective breeding or 'artificial selection'. In 1859, Darwin published *On the Origin of Species by Means of Natural Selection, or the Preservation of Favoured Races in the Struggle for Life.*

Fitness, filters and fate

'Survival of the fittest' is how many people understand natural selection. The expression was coined by the philosopher Herbert Spencer in 1864, and Wallace is partly to blame for its popularity. He disliked 'natural selection' as the phrase might be taken literally to imply a conscious 'selector' rather than mindless Mother Nature. After some nagging, Darwin replaced his phrase in the book's fifth edition (1869). Wallace then went through his own copy, crossing-out every 'natural selection' and inserting 'survival of the fittest' by hand. Natural selection was best summarized by Darwin: 'multiply, vary, let the strongest live and the weakest die'. The last part is survival and 'multiply' is reproduction, but how do organisms 'vary'? Since the modern evolutionary synthesis of the 1930s, when Mendel's laws of inheritance were combined with natural selection, biologists have known that the main source of variety is mutation, which creates individuals with combinations of gene variants. Each 'genotype' determines a 'phenotype', the visible effects that influence an organism's fitness – its ability to survive and reproduce.

> The objections now made to Darwin's theory apply, solely, to the particular means by which the change of species has been brought about, not to the fact of that change
>
> Alfred Russel Wallace

Imagine selection as a succession of filters that influence the fate of a new mutation in the gene pool. If the mutation raises fitness – such as a variant that protects a plant against drought – then it passes through each filter and spreads through the population by 'positive selection'. Good mutations enable a species to adapt so this is also known as 'Darwinian selection'. If the mutation reduces fitness (worst case: lethal), then it can be stopped in its tracks by 'negative selection'. Bad mutations are weeded out from the population so this is called 'purifying selection'. If a mutation is both beneficial and harmful, it can be maintained by 'balancing selection'. One example is the gene variant that causes sickle cell trait, where one copy protects against malaria but mutations on both chromosomes cause disease.

Sexual selection

Natural selection is also classified by what does the selecting. 'Sexual selection' occurs through choice of mates, while 'ecological selection' is pressure from any other part of the environment. Darwin saw sexual selection as distinct, but modern biologists consider it a subset of natural selection. Sexual selection is one example where Wallace and Darwin differed: Wallace believed that female birds were duller to protect them from predators; Darwin believed males are brightly coloured to attract females. Although the discoverers of natural selection disagreed on a few things, they remained friends. As Darwin wrote in a letter to Wallace in 1870: 'I hope it is a satisfaction to you to reflect – and very few things in my life have been more satisfactory to me – that we have never felt any jealousy towards each other, though in one sense, rivals.'

The condensed idea
Species continually adapt to a changing environment

45 Genetic drift

Natural selection drives evolution forward, but it is not the only force that causes a population to change over time. When individuals survive and reproduce through luck, their DNA can be lost or spread after drifting through the gene pool.

Whereas natural selection was peacefully discovered by Darwin and Wallace, the theory of genetic drift arose from a conflict between two mathematical geniuses: Ronald Fisher and Sewall Wright. Fisher was born in London and showed early promise with numbers, but was very short-sighted, developing extraordinary mental arithmetic to compensate. Wright grew up in Illinois, his father a former economist and polymath nicknamed the 'Illinois Prairie Leonardo'. The precocious young Wright could calculate cube roots before he even started school.

In the 1930s, Fisher, Wright and J.B.S. Haldane established the field of population genetics, foundation of the 'modern evolutionary synthesis' that combined natural selection with Mendel's laws of inheritance. Although Fisher and Wright agreed on the main mechanism (selection drives species to adapt) they argued over details, most notably how evolution creates novelty. Fisher believed it occurs faster by mixing all members of a population, Wright proposed the 'shifting balance' theory: new gene combinations and novel features originate faster via migration between partially isolated subpopulations. At the heart of disagreements was the role of random chance. Fisher said it played a small part; Wright thought it was important.

> Calculating the rate of evolution in terms of nucleotide substitutions seems to give a value so high that many of the mutations involved must be neutral ones.
>
> Motoo Kimura

Choice, chance and change

Imagine the robot uprising has finally happened and your master is a machine that plays games with beans. It drops dozens of red kidney beans and white cannellini beans into a bowl, and gives you a moment to choose ten. Being a fan of chilli con carne, you grab a handful of mostly red ones. The machine explains that after growing your beans into plants, you will harvest their beans for a fresh bowl. This repeats for successive generations: the machine occasionally adds its own beans representing mutations, while you act as natural selection, trying to grab red every time.

The machine gets bored. Instead of choosing beans from a bowl, you have to pick blindly from a bag, which gives 10 possible combinations of red and white beans. From an average ratio that initially hovers around 5:5, you get more of the same colour over time (6:4, 7:3, 8:2, 9:1, 10:0) because the beans you pick are grown from plants of matching colour. So with a 9:1 ratio of red to white in a bag of 100 beans, there is a remote possibility of picking 10 whites, but 10 reds is statistically more likely. The beans represent alternative variants of genes (alleles) that become more or less common due to random sampling from the bag.

Genetic drift is the fluctuation in allele frequency over time, due to random sampling. Peter Buri illustrated it in 1956 by creating over 100 populations of fruit fly. His insects carried alleles for red or white eyes, with an initial frequency of 0.5, meaning that 50 per cent of

Fluctuating allele frequency

Genetic drift occurs when an allele (gene variant) becomes more or less common in a population over time, due to random sampling at each generation. Starting from a frequency of 0.5, where half of individuals carry an allele, it can either: fluctuate around an average frequency (middle line); be inherited by every member of a population (top); or drift out of the gene pool (bottom).

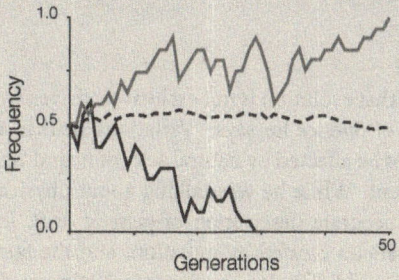

Bottlenecks

When the population size shrinks, a random sample of remaining individuals gives less variation. One consequence of this low genetic diversity is less raw material for natural selection – less chance of lucky individuals with mutations that might allow the population to adapt to environmental change. This can tip an endangered species towards extinction. Bottlenecks can also lead to speciation through the 'founder effect', proposed by evolutionary biologist Ernst Mayr in 1942. Mayr called the mathematical work of Fisher, Wright and Haldane 'beanbag genetics', so it is ironic that the founder effect is down to genetic drift: the colonizing individuals carry a small sample of alleles (gene variants) from the original population, so some alleles can be lost or spread by chance. One example in humans occurs among the Afrikaner population, for Huntingdon's disease. The genetic disorder of the nervous system is normally rare, but unusually frequent among white South Africans because one of the Dutch colonists who settled there in 1652 was unlucky enough to carry the disease-causing allele. The allele can be missed by natural selection because people reproduce and pass it on before they even realize they have it.

alleles in each population being red. He bred flies for 19 generations, randomly picking eight males and eight females each time, irrespective of eye colour. By the end, one-quarter of populations had lost the red allele, another quarter all had red eyes, while half still had an allele frequency of 0.5.

Neutral theory

Darwin was aware that evolution is not exclusively driven by selection. In *On the Origin of Species* he says: 'Variations neither useful nor injurious would not be affected by natural selection, and would be left a fluctuating element'. While he was talking about physical features, this is a spookily accurate description of genetic drift. In genetics, variations are the alleles created by mutation, and the fate of a new mutation is determined by choice or chance (selection or drift). Both can cause a mutation to be lost from the gene pool or spreads until it is carried by every individual ('fixed' within a species).

The fate of a mutation is influenced by how it affects fitness, an organism's ability to survive or reproduce. While we often say mutations are good or bad, they can also be neutral. Until the 1960s, many researchers assumed that selection would push most good mutations through a gene pool. Then sequencing technology enabled scientists to read the letters in proteins and later DNA, allowing them to compare the same molecules in different species and count the number of differences. In 1968, Japanese geneticist Motoo Kimura used this data to calculate the nucleotide substitutions (single-letter replacements) in the genomes of humans and fruit flies. He found so many mutations that it seemed unlikely they were all chosen, suggesting they had accumulated through random genetic drift.

Kimura's neutral theory of molecular evolution was refined in 1973 by Tomoko Ohta, who suggested that even slightly harmful mutations – those with a small impact on fitness – were also ignored by selection.

Population size

The demanding bean machine has now decided you will only pick four beans from the bag, giving red:white ratios of 0:4, 1:3, 2:2, 3:1 or 4:0. Through random sampling, it takes less time to drift towards an all-red or all-white population. If the machine goes back to letting you choose beans from a bowl, but you only get to quickly grab four rather than ten, it is more likely you will accidentally have a handful of white beans. The strength of selection is influenced by population size. Small populations are more susceptible to random sampling and genetic drift, which can reduce genetic variation by creating a bottleneck. The neutral theory explains the fate of mutations that are 'invisible' because they have little to no impact on fitness, or because a small population leads to random sampling and genetic drift.

The condensed idea
Evolution driven by random chance

46 The selfish gene

Survival of the fittest individuals is one way to view natural selection – another is from the perspective of genes, molecules that selfishly use the organisms carrying them to replicate and pass from one individual to the next generation. This view of evolution helps to explain a selfless social behaviour: altruism.

Published in 1976, Richard Dawkins' *The Selfish Gene* brought together the work of 20th-century biologists who had developed a 'gene-centred view of evolution'. This allows us to think of natural selection as a process that forces variants in the gene pool to sink or swim, behaving as if they are competing against each other, so surviving variants reproduce through the individuals carrying them. Selection acts on an individual and its genes at the same time, like a racing car and its components. Dawkins, who had studied animal behaviour, decided to focus his book on how the gene-centred view explains altruism.

Why do we help others? Why should any organism help another? Social behaviours occur in four categories based on costs and benefits to an actor (individual performing an act) and the recipient: mutual benefit helps both parties, spite harms both, selfishness benefits the actor and altruism is costly. Altruists will spend resources, from time and food to the ultimate cost: self-sacrifice. Kindness and charity involve a concern for the welfare of other people, but touchy-feely explanations do not apply to mindless animals.

Group selection

Natural selection is 'survival of the fittest', but the fittest what? Genes? Individuals? Groups? Scientists once assumed that nature could weed out unfit entities at any level, so altruism can be explained by organisms acting 'for the good of the group' or 'benefit of the species'. For example, zoologist V.C. Wynne-Edwards claimed in 1962 that individuals deliberately limit their birth rate to minimize the burden on a population. But around the mid-20th century, evolutionary biologists like George Williams argued against naïve group selection. One problem is that a group of altruists is vulnerable to invasion by cheaters who reap the benefits of cooperation without the costs, and

Adoption

Caring for another parent's offspring as your own – adoption – has been reported in over 60 mammal species. As an example of altruism, it can be used to test Hamilton's inclusive fitness theory, which would predict that surrogate mothers will be more likely to adopt orphans that are closely related. Results from the wild are often unclear, however, as it is hard to calculate the fitness cost to adopters that live in groups, as altruistic behaviour can have benefits unrelated to inclusive fitness, like cementing social relationships. One way around this is to study solitary 'asocial' species like red squirrels. In 2010, Canadian ecologist Jamie Gorell and colleagues surveyed 19 years' worth of data at one site from 2,230 squirrel litters, and found that orphans without nearby kin were never adopted. The researchers identified only five adoptions and, based on known family trees and genetic tests, they showed that orphans were always related to surrogate parents by at least 12.5 per cent – the equivalent of being first cousins. Hamilton's rule can therefore explain occasional altruism among animals that are not naturally social.

so they save resources. The three fathers of population genetics – Ronald Fisher, Sewall Wright and J.B.S. Haldane – made similar arguments with gene pools. Altruism needed a better theory.

Kin selection

Why do we care for our children? It costs time and money, for clothes and education, yet people say parenting is rewarding. So what are the benefits? Darwin hinted at the explanation in *On the Origin of Species*, after noting an exception to the need to reproduce for natural selection: insects such as ants, bees and termites can form societies that include sterile workers. Darwin said this 'difficulty' disappears by remembering that 'selection may be applied to the family, as well as to the individual.' In 1964, biologist John Maynard Smith dubbed this 'kin selection'.

When asked whether he would risk his life to save a drowning brother, J.B.S. Haldane once quipped, 'No, but I would to save two brothers or eight cousins.' In 1955, he explained this nepotism in terms of a hypothetical gene that influences the behaviour of jumping into a river. Based on random assortment of genes when parents pass chromosomes to offspring, you share 50 per cent of DNA with siblings, while cousins are 12.5 per cent related (so eight gives 100 per cent). Kin selection explains why 'blood is thicker than water'.

Inclusive fitness

Altruism is a perplexing problem in terms of natural selection because the cost reduces an individual's fitness to survive and reproduce – selection should only favour altruism if it raises individual fitness. The benefit that helps balance cost is the amount of relatedness or 'inclusive fitness', as proposed by British evolutionary biologist W.D. Hamilton in 1964. Hamilton's models essentially asked: When do the benefits to a relative outweigh the costs to myself? The models can be simplified to an equation now called Hamilton's rule: $rb > c$, where 'c' is cost to an actor's inclusive fitness, 'b' is benefit and 'r' is relatedness to a recipient (the same genes). Because relatedness is multiplied by benefits ($r \times b$), altruism is more likely when actor and recipient are close relatives.

> We are survival machines – robot vehicles blindly programmed to preserve the selfish molecules known as genes.
> Richard Dawkins

Hamilton tested inclusive fitness on Darwin's 'difficulty': the eusocial insects. Labour is usually divided into one egg-laying queen plus hundreds, thousands or millions of sterile workers, each doing jobs like foraging for food and rearing young. Whereas humans are 'diploid', with one set of genes from each parent (so a female is 50 per cent similar to a parent or sibling), social insects are typically 'haplodiploid' because the queen manipulates fertilization so that sisters are 75 per cent related (only 25 per cent to brothers). Hamilton found that inclusive fitness could explain insect eusociality, but his equations had flaws. In 1970, troubled genius George Price provided a new formula to explain various natural phenomena – the Price Equation – which helped Hamilton refine his work.

Kin selection raises inclusive fitness by enabling an individual to influence whether its genes are passed to the next generation by directing altruism towards relatives with the same genes. How does that actually work? Hamilton originally proposed two ways. Limited dispersal occurs when individuals live in the same area, so altruistic behaviour could evolve because individuals expect one another to be related. The other way – discrimination – causes complications, as illustrated by a thought experiment from *The Selfish Gene*: if relatives can recognize one another through a conspicuous green beard, the strategy initially works but soon becomes vulnerable to invasion by cheaters, or 'false beards'. Altruists would also end up selecting individuals carrying the 'beard gene' (and linked DNA) itself – not relatedness.

The condensed idea
An evolutionary perspective helps explain altruism

47 Cooperation

Kin selection, helping those who share the same genes, explains altruism among relatives. But cooperative behaviour also occurs in unrelated individuals – and even between members of entirely different species. How could altruism evolve in such cases? One explanation is mutual benefit through 'reciprocal altruism'.

In nature, altruism is most common among relatives. The ability to survive is influenced by your resources, so spending them on other individuals incurs a 'fitness cost' to yourself. Biologist W.D. Hamilton did the maths to show that those costs can be offset if the individual you help is a relative – because they share a fraction of your genes, in a way you are also helping yourself. Hamilton's 'inclusive fitness' idea was popularized in Richard Dawkins' 1976 book *The Selfish Gene*, which also describes the dominant theory to explain cooperation among non-relatives.

Reciprocal altruism

'You scratch my back, and I'll scratch yours.' This is the basic idea behind reciprocal altruism, the theory proposed in 1971 by American sociobiologist Robert Trivers, who said that the costs and benefits of cooperative behaviour are fundamentally altered when a social interaction is not simply a one-off situation, but the actor (performing the altruistic act) and recipient (the one who benefits from it) are likely to meet one another again.

One very obvious example of reciprocal back-scratching is grooming among primates. A less literal example occurs between cleaner fish and their hosts, a symbiotic relationship that could be called 'mutualism' as it increases the fitness of both species: hosts benefit from parasite removal; cleaners get a meal. Other aspects of the relationship do seem altruistic, however. Some hosts allow cleaners to enter their mouths without eating them, and even chase away threats to cleaners, expending energy and risking harm – a fitness cost – to themselves.

Exchanging favours

Trivers' theory suggests that cooperation can evolve through a positive feedback loop: after one actor does an initial 'favour' for a recipient, the other reciprocates by offering a 'favour in return' (the recipient becomes an actor). If repeated interactions provide fitness benefits, genes associated with the cooperative behaviour will spread through a population. But as with natural selection favouring any behaviour within a group, cooperation creates an evolutionary conundrum: what prevents an individual from cheating, taking the benefits without the costs? What stops one host from snacking on its cleaner fish while everyone else exercises self-restraint? Such cheaters save resources to spend on other things that increase fitness, so their selfish genes – and behaviour – should become common after spreading through the host gene pool.

A more accurate summary of reciprocal altruism is: 'You do something for me and I'll return the favour – later.' The later part is important because it introduces a delay between favours, giving altruistic individuals the time to spot cheaters within a group, to discriminate and even punish them – either by not repeating the cooperation or through active aggression. In a group populated by

Genetic conflict

Cooperation may be common among relatives, but even the most closely-related individuals – parents and offspring – can come into conflict. Like so many of nature's disagreements, this is ultimately driven by a fight over limited resources. The idea builds upon W.D. Hamilton's maths for 'inclusive fitness' due to genetic relatedness: offspring share 50 per cent of genes with each parent, but only share *up to* 50 per cent with brothers and sisters. As a consequence, offspring will selfishly try to benefit themselves at the cost of siblings, while parents aim to distribute resources equally among current and future progeny. In 1974, Robert Trivers argued that this causes disagreements over how much parents should invest in individual offspring. In the 1990s, Trivers looked at the conflict between an organism's genome and its selfish genetic elements, parasites that can be harmful to their host and contribute to junk DNA. This led to a comprehensive 2008 book co-authored with evolutionary geneticist Austin Burt, *Genes in Conflict*.

such savvy altruists, the 'gene for cheating' would not be favoured by selection. From this, Trivers suggested that reciprocal altruism has a wider impact on social behaviour, explaining things like gratitude, trust and suspicion, as well as guilt and attempts to 'make amends'. Biological interactions thus provide a path for the evolution of fairness and justice.

Play to win

The reciprocal altruism theory has been backed up by research in the field of game theory, the study of strategy and decision making, whose applications extend beyond biology, everything from economics to nuclear deterrence. Its most famous thought experiment is 'the prisoner's dilemma', where two players are given the opportunity to cooperate for a small shared benefit, or gamble on a significant individual reward – or a penalty if they are caught out – by pinning a crime on the other player.

In 1980, political scientist Robert Axelrod organized a tournament of repeated games of the prisoner's dilemma. Each entry was an algorithm – a set of instructions for how a player should instinctively play, and how they should react to the other player's actions. Entries were scored based on the outcome over 200 computerized iterations of the game. One strategy, proposed by mathematician Anatol Rapoport, consistently came out on top: a simple algorithm in which a player repeatedly chooses cooperative behaviour unless the other player defects, at which point they punish the other player by defecting in the next game, and only return to cooperation if the traitor sees the error of their ways. Similarities between Rapoport's 'tit-for-tat' algorithm and reciprocal altruism inspired Axelrod to collaborate with W.D. Hamilton, the discoverer of 'inclusive fitness', combining natural selection and game theory in a 1981 paper entitled 'The Evolution of Cooperation'.

> Fairness ... does not flow from kinship – justice has biological roots.
> Robert Trivers

Mutualism or manipulation?

For several decades, reciprocal altruism has been widely accepted as the likely explanation for cooperation among non-relatives. However, some biologists have since suggested that – apart from humans –

animals do not commonly exchange resources or services. In a 2009 review, 'British zoologist Tim Clutton-Brock examined some classic examples and found that none met his strict criteria to demonstrate they were definitely altruistic, concluding that many examples of cooperation between non-kin are probably cases of mutualism or manipulation.

Is cooperation rare in nature? This question is hard to answer, partly because behaviour can be interpreted in different ways, as illustrated by the textbook case of bloodsucking bats. In 1984, American sociobiologist Gerald Wilkinson reported that when vampire bats return home after hunting, successful animals would sometimes regurgitate food for hungry roost-mates, and this altruism was often directed towards relatives. Theoretical biologist Peter Hammerstein proposed that food sharing with non-relatives is a by-product of kin selection – recognition gone wrong, or 'miscalibrated kin recognition hypothesis' – whereas Tim Clutton-Brock suggested that sharing is simply manipulation by others in the group, with persistent begging by unfed bats leading to coercion, the 'harassment hypothesis'. But in 2013, Wilkinson did a two-year experiment to test these hypotheses. After fasting 20 vampire bats and providing food every 48 days, he showed that two thirds of sharing occurred between unrelated pairs of individuals (which seems high for kin-selection-gone-wrong) and food donors initiated food sharing more often than recipients (inconsistent with harassment). At least in vampire bats, cooperation provides mutual benefits to fitness. Even blood-suckers are not too selfish to share.

The condensed idea
Play nice or suffer the consequences

48 Speciation

Darwin's *On the Origin of Species* focuses on the evolutionary mechanism that produces a tree of life, without really explaining how one branch becomes two. Although termed a speciation 'event', this process is typically a slow, gradual separation that ends with members of a population no longer able to interbreed.

Before we can answer the question of how new species form, we must first ask: 'What is a species?' Taxonomists group organisms by shared features such as morphology, but many biologists prefer the 'biological species concept', popularized by German-American evolutionist and ornithologist Ernst Mayr. In a 1942 book *Systematics and the Origin of Species*, Mayr defined species simply as populations that can interbreed.

There are two major routes to new species in animals, plants and other organisms that reproduce through sex. Sympatric speciation occurs after genetically distinct individuals emerge from within a population, eventually forming two non-interbreeding species with overlapping ranges. One example of this is cichlid fish in central Africa's great lakes, where mate choice by females has driven sexual selection and created distinct groups. The second major route to new species is allopatric speciation – splitting a population with a geographical barrier, which is thought to trigger most speciation events. While there are other routes to new species, it is often difficult to prove that two groups have not been geographically separated at some point in the past.

Geographical isolation

Allopatric speciation begins when a barrier – perhaps a rising mountain range, a glacier field or even a new road – physically splits a population in two. The barrier does not need to be permanent – it only needs to last long enough to kickstart the process of creating two 'incipient' species. Even migration can trigger allopatric speciation, if some members of a population manage to cross a barrier they would not normally be able to overcome, enabling them to fill a new, vacant ecological niche. Darwin's finches, studied for over 40 years by evolutionary biologists Peter and Rosemary Grant, are a famous example: more than a dozen separate species are scattered across the young islands of the Galápagos archipelago, despite its isolation from mainland South America. In 2001, the Grants worked with geneticists to show that the finches are descended from grassquits, songbirds from Central and South America. It is thought that around 2.3 million years ago in the last ice age, ancestors of the island species managed to hop across ice from the mainland.

Artificial selection in the lab has also provided evidence for allopatric speciation. In 1989, American biologist Diane Dodd divided one population of *Drosophila pseudoobscura* fruit flies into two groups to simulate geographic isolation, breeding half on a maltose diet while the other half lived on starchy food. When brought back together after a year, the maltose group preferred to mate with other 'maltose flies', while the starch group preferred 'starch flies'. Separation had changed their sexual behaviour – a geographic barrier had created a reproductive barrier.

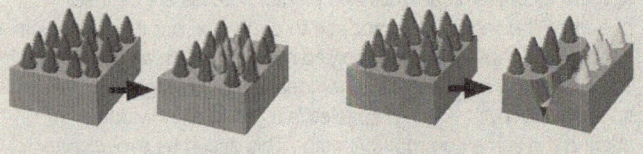

Two major routes to speciation: sympatric speciation (left) occurs after genetically distinct individuals emerge from within a population and stop interbreeding, forming two species with overlapping ranges, whereas allopatric speciation (right) occurs after a geographic barrier physically separates the population.

Reproductive isolation

What prevents reunited incipient species from interbreeding to recreate a single population? Speciation continues if barriers before or after mating create reproductive isolation. One pre-mating factor is the ability to recognize members of your own species through sights and sounds: in the 1980s, for example, the Grants observed that male finches would only approach a loudspeaker playing songs from their own species.

Hybrids are formed when members of two populations mate. If their offspring are healthy and not sterile, their parents were *not* from different species. This is the heart of the biological species concept, first proposed by Russian population geneticist Theodosius Dobzhansky in 1935. Dobzhansky also developed the idea of a 'gene pool', to which every individual contributes. Speciation is the result of barriers that block gene flow in a population. One way this can happen is by changing the number of chromosomes: most individuals have a pair inherited from each parent, but some carry multiple copies, making them 'polyploid'. This is common in plants, but not in animals. In 1937, Dobzhansky suggested that reproductive barriers could be created by the accumulation of incompatible genetic variants. Hermann Müller, the first person to

Species are groups of actually or potentially interbreeding natural populations, which are reproductively isolated from other such groups.

Ernst Mayr

induce mutations in fruit flies, came to a similar conclusion in 1942. Their theory, the Dobzhansky–Müller model of hybrid incompatibility, was proven right by American geneticists Jerry Coyne and Allen Orr in the 1980s. After crossing two related fruit flies, Coyne and Orr identified several genetic factors that influence whether offspring would be fertile.

The Wallace and founder effects

British naturalist Alfred Russel Wallace not only proposed natural selection with Darwin, but also founded the field of biogeography (the study of how species are distributed) and contributed to the theory of speciation. In his 1889 book, *Darwinism*, Wallace suggested that once two populations have diverged to a point where each is well-adapted to its environment, any hybrids would be less adapted, causing them to be weeded out by natural selection. This process reinforces the differences between the two incipient species late in the process, and is now known as the Wallace effect.

Reproductive isolation can also be driven by chance, through genetic drift. Individuals that migrate to a new environment, such as the ancestors of Darwin's finches that reached the Galápagos islands, are a subset of a larger population. This small, random sampling initially contains few variants in the gene pool, limiting their early evolution. This concept, the 'founder effect', was proposed by Ernst Mayr in 1942.

Whatever the processes that drive isolation, the end result is that reproductive systems of two species become incompatible, preventing fertilization because eggs no longer recognize sperm. Later, sexual organs no longer fit together properly. As the two groups accumulate more and more differences over time, their two branches on the tree of life grow further apart, to the point where even non-biologists can distinguish between distinct species.

The condensed idea
New species are formed by barriers

49 Extinction

While our planet's current collection of organisms might seem impressive, it is but a fraction of the biodiversity over time: over 99 per cent of species have disappeared during the history of life on Earth. Most went extinct slowly, but some species have died out relatively rapidly, due to mass extinction.

Extinction is a natural phenomenon. Given the rate at which species are currently dying out from human activity, this is easy to forget. Whatever the cause, when the number of deaths over time outpaces the birth rate, a population is doomed to go extinct. The official moment of extinction occurs when the last living member of a group passes away, but a population might be effectively extinct long before then, because there are too few individuals to reproduce. As the group shrinks, there is more inbreeding, creating a shallow gene pool containing too little variety. Organisms are in poor genetic health and will be less likely to survive natural selection, dying of disease or being outcompeted by other organisms.

Species extinction

Whether triggered by natural or artificial changes, the ultimate cause of extinction is always the same: a species fails to adapt to its environment. Invasive species can emerge by migration or after being introduced by humans, for example, and drought can be due to normal weather cycles or anthropogenic climate change. Species can also go extinct via a knock-on effect, if something they depend upon disappears from their ecosystem: predators lose prey, parasites lose hosts, plants lose bees. Extinction can be a consequence of speciation, too: as one branch of the tree of life splits into two, the ancestor of the two descendants goes extinct. Just as new kinds of organisms appear via speciation, others disappear through extinction.

Scientists did not always think species could die out. Until the late 18th century, many assumed fossils in rock were the remains of living creatures; others believed God would never allow His creations to vanish from the face of the Earth. If an organism could not be found, it had moved or was alive somewhere else. The first clear evidence of extinction was presented in 1796 by anatomist Georges Cuvier, who described his work on fossilized bones to the French Academy of Sciences. He claimed that African and Indian elephants were distinct, and that the mammoths and mastodons of Europe and Siberia were 'lost species'. Cuvier, however, did not believe in evolution through gradual 'transmutation' of species. He thought new species appeared following sudden 'revolutions', periodic catastrophes that wiped out numerous species. Although wrong about evolution, Cuvier was right about the catastrophes.

Mass extinctions

While the majority of organisms have disappeared via the slow, continuous extinction of individual species, occasional natural disasters can affect ecosystems worldwide and kill a massive number of species in a relatively short period of time. Palaeontologists recognize five events when the biodiversity loss was so huge that they constitute a mass extinction. The 'Big Five' extinctions were identified in 1982 by Americans David Raup and Jack Sepkoski, who calculated extinction rates among 3,300 families of marine vertebrates and invertebrates. Starting from when animals first appear in the fossil record 542 million years ago (mya), they noticed a sharp drop in the number of families at five points. These are now dated to the ends of the Ordovician, Devonian, Permian, Triassic and Cretaceous geological periods (444, 372, 251, 201 and 66 MYA). The most massive was the end-Permian event or 'Great Dying' – the only time when forests and coral reefs virtually disappeared, and family diversity was cut in half. The most recent event was the end-Cretaceous extinction 66 MYA, which reduced families by 11 per cent and wiped out non-avian dinosaurs.

Mass extinctions can be caused by dramatic climate change, set off by external factors such as meteorite impacts or internal forces like volcanic activity, which lead to global warming or cooling, creating a 'Greenhouse Earth' or 'Snowball Earth'. Such changes happen too fast for feedback mechanisms like rocks or plants to compensate quickly enough, altering the environment beyond the ability of many organisms to adapt. Life rebounds after a mass extinction, as survivors evolve to occupy vacant ecological niches, so extinction can be a creative process.

The sixth extinction

Conservationists warn that we are in the midst of a sixth mass extinction, driven by human activity. The Worldwide Fund for Nature states that the two primary threats are habitat loss and degradation, and exploitation through hunting and fishing. The biodiversity crisis is named after the current geological epoch, the 'Holocene extinction'. Its most famous victim is the dodo, a large flightless bird last seen in 1662: when the Dutch settled on the island of Mauritius, they destroyed the bird's forest habitat and introduced mammals that competed for food. According to biologist Rodolfo Dirzo, 322 species of terrestrial vertebrates have gone extinct since 1500. Among remaining species, there has been a 25 per cent decline in land vertebrate populations and a 45 per cent decline among invertebrates.

> For an evolutionary biologist to ignore extinction is probably as foolhardy as for a demographer to ignore mortality.
>
> David Raup

From a palaeontological point of view, the current biodiversity crisis does not yet qualify as a mass extinction, but this is not a fair comparison because 'Big Five' figures were calculated from fossils, whereas living organisms might still be on their path to extinction. Assuming species classed as 'threatened' by the International Union for Conservation of Nature have passed the point of no return, the extinction magnitude is around 23 per cent. Based on a hypothetical scenario by palaeobiologist Anthony Barnosky, where all currently threatened species go extinct within a century, we could reach the 75 per cent level in around 300 years.

Another method to test whether we are living through the sixth mass extinction is to compare the current rate of species loss against a natural background rate, estimated at 0.1 extinctions per million species per year. In 2014, conservation ecologist Stuart Pimm calculated that the current rate is 1000 times higher than the background rate.

De-extinction

Could scientists resurrect extinct species using the cloning technology that created Dolly the sheep? This may sound like science fiction, but the feat was achieved in 1999, when Spanish biologists cloned a mountain goat (Pyrenean ibex) from frozen cells. So animals whose tissue has previously been stored can be brought back from the dead. Things are less promising for species that have already died out, however, as cloning requires viable genetic material and DNA quickly breaks down – especially in warm temperatures – meaning that producing the dodo or thylacine (Tasmanian tiger) is probably impossible. And while Siberia's cold conditions may well preserve DNA from a woolly mammoth, even that material will be fragmented, with too many missing pieces. So-called 'de-extinction' actually involves editing the genes of a living species to add features from an extinct relative – creating a woolly elephant, for example. DNA from such genome engineering would be used to make stem cells for an egg that could be implanted into a surrogate mother or artificial womb.

The condensed idea
Species fail to adapt to environmental change

50 Synthetic biology

On 20 May 2010, American geneticist J. Craig Venter announced that his team had created the world's first synthetic cell. The microbe represents the next step in genetic engineering: instead of modifying an existing organism's DNA, a genome is built from scratch – an achievement that will lead to designer life forms.

Venter is famous for leading the private firm that raced against the public Human Genome Project to read our complete DNA sequence. The race ended in a draw in 2000, and Venter calls it a 'three-year diversion' from his main goal: synthetic life. In 1995, working with microbiologists Clyde Hutchison and Hamilton Smith, his team was the first to read the genome of a free-living organism, the bacterium *Haemophilus influenzae*, and the smallest genome, *Mycoplasma genitalium*. In 2003, they managed to write a genome, using synthetic nucleotides to recreate the 5,000-letter sequence of the Phi X174 phage virus. Then, in 2010, they did the same for *Mycoplasma mycoides*, making a synthetic cell from a digital code of one million letters in a computer. The media nicknamed it 'Synthia'.

Building blocks

The main aim of synthetic biology is to build living machines. Researchers are compiling a set of standardized parts, genetic Lego blocks called BioBricks, to make it easy to mix and match different components. Thousands of modules have been uploaded to the Registry of Standard Biological Parts, a database hosted by the Massachusetts Institute of Technology. Designer organisms could combine parts with distinct functions, like sensors to detect chemicals such as toxins and indicators to reveal their presence. In 2009, for instance, Cambridge University students won the annual international Genetically Engineered Machines (iGEM) competition by making seven strains of *E. coli* that produced different coloured pigments, dubbed 'E. chromi'.

One key component of a machine is an on/off button. In 2000, bioengineer James Collins created a toggle switch in *E. coli* that can be flipped between two states. The switch consisted of two genes, each encoding a 'repressor' protein that blocked the activity of the other so that when one gene was 'on', the other was 'off'. It can be controlled by giving the cell a specific chemical or altering temperature, and could potentially be used to activate other genes.

Minimal genomes

Machines need a chassis to support their parts. Every genome includes DNA that is necessary for the running and maintenance of a cell – 'housekeeping' genes encoding proteins and RNA that are needed for vital metabolic reactions and cell division, for example. But these genes are not modular units like BioBricks, making them hard to isolate. One way to design a chassis is to painstakingly delete each gene, one at a time, and observe whether an organism survives. Those that are not essential can then be removed, leaving the chassis or 'minimal genome'. This is what Venter's team did with *Mycoplasma genitalium*, finding that of the bacterium's 482 genes, around 100 are not essential for survival, at least in lab conditions. Removing non-essential genes helps prevent their products causing complications when new genes are added.

The chassis can also be streamlined. In 2009, a team led by George Church invented a technique called MAGE (Multiplex Automated Genome Engineering) to speed up the creation of designer organisms. In 2011, the team of geneticists used MAGE to re-engineer *E. coli* genomes by performing the equivalent of find-and-replace in a text document, changing every instance of the three-letter word 'TAG' to 'TAA' in the DNA. When the gene-reading machinery encounters those two words, both translate as 'stop' (the end of instructions for making an encoded protein) but switching to using 'TAA' alone means the translator for 'TAG' is redundant and can be deleted. This would protect a reengineered *E. coli* from viruses that rely on 'TAG' when replicating in a host bacterial cell.

Safety and security

The potential of synthetic biology is both exciting and scary. One fear is that it could be used for bioterrorism: virologists recreated poliovirus by synthesizing its genome in 2002, while in 2005 researchers at the US Centers for Disease Control and Prevention resurrected the virus behind the 1918 Spanish flu pandemic. A related concern echoes old arguments against genetically modified (GM) organisms: containment to prevent an organism escaping into the wild. As a result, synthetic biologists perform extensive risk assessment and have a code of ethics and responsibility on top of the usual regulation for working with potential biohazards. These include safeguards to stop organisms from surviving outside the lab. Venter, for example, has deleted genes so his bacteria cannot grow without specific nutrients, which puts a microbe on a short leash. James Collins, who developed the on/off toggle switch, is developing a genetic 'kill switch' that triggers production of toxic proteins in response to certain chemicals. In future, synthetic life could be built from molecules not found in nature, blocking replication.

> **This is the first self-replicating species that we've had on the planet whose parent is a computer.**
> J. Craig Venter

Digital designs

Venter hopes to create algae that absorb carbon dioxide from the air and release biofuels. He also has a vision for a machine that creates custom organisms on demand from a digital DNA code, growing cells to provide an insulin prescription or a vaccine against a pandemic. Venter calls this machine a 'digital biological converter' – it sounds like science fiction, but there is already a basic prototype.

What is life? Synthetic biology raises numerous philosophical questions about relationships between populations of organisms, and especially between man and everything else. Although many populations can influence the evolution of others to some degree – even driving speciation and extinction – humans are the only species with the ability to create life from scratch. Is a living thing still 'natural' if it was made by an artificial process? The only thing we know for certain is that if an organism has life's features and does life's processes, it is life.

Genome editing

CRISPR has enabled a revolution in genetic engineering. For half a century, scientists used viruses or bacterial enzymes to modify DNA or introduce genes into a genome – approaches that were often hit-or-miss, causing undesirable or dangerous 'off-target' effects. By contrast, CRISPR uses molecular scissors to snip specific letters in the double helix (a cell's repair machinery fixes the cut). This technology was developed from an adaptive immune system in microbes that includes a DNA sequence or 'CRISPR' (Clustered Regularly Interspaced Short Palindromic Repeats) and the enzyme Cas9 (CRISPR-associated protein 9), which is guided to a DNA target by two RNA molecules. But in 2012, Emmanuelle Charpentier and Jennifer Doudna invented a combined molecule, single guide RNA, that makes it easy for researchers to design sequences that match a DNA target – a precise tool for genome editing. CRISPR is used to create GM organisms and in human gene therapy to correct mutations that cause conditions such as sickle cell disease. Controversially, CRISPR has even produced gene-edited babies in which the CCR5 protein, which lets HIV invade cells, was disabled to protect against infection.

The condensed idea
Life by design

Glossary

Adaptation Evolution that produces characteristics that provide fitness in an environment.

Alleles Different variants of a gene.

Amino acids The chemical building blocks of polypeptides.

Carbohydrates Molecules made of carbon, hydrogen and oxygen that fuel most organisms.

Cells Units of life with fatty membranes to separate metabolism from the surroundings.

Characteristics Physical features of an organism, including biochemistry invisible to the naked eye.

Chromosomes Structures made of DNA and associated proteins.

Cytoplasm The contents of a cell, excluding the nucleus, in a watery solution (cytosol).

DNA Deoxyribose nucleic acid, a molecule with four bases (A, C, G, T) that forms paired strands in the double helix.

Embryo A multicellular organism during early development, between fertilization and birth.

Environment The physical surroundings and biological organisms within an ecological habitat.

Enzymes Proteins or RNA molecules that enable biochemical reactions (catalysis).

Epigenetics The transmission of biological information that is not encoded as a sequence of letters in DNA.

Eukaryotes Organisms consisting of one or more complex cells that usually contain a nucleus.

Evolution Change to a population over time.

Fitness The ability to survive and reproduce.

Gametes Reproductive cells, often egg or sperm.

Gene expression Conversion of biological information into physical characteristics.

Gene pool The collection of alleles in a population.

Genes Units of heredity that encode instructions for making a protein or RNA molecule.

Genetic code The rules that allow genetic instructions to be translated into proteins.

Genetic drift The process that causes evolution through random sampling from a gene pool.

Genetic material *see* Nucleic acids.

Genome The full set of genes or nucleic acid within a cell, individual or species.

Genotype The combination of alleles for one or more genes.

Habitat A population's natural home.

Heredity Transmission of genes between generations, often interpreted as parents to offspring, but also by cell division.

Homeostasis Keeping internal conditions relatively constant.

Inheritance *see* Heredity.

Metabolism The biochemical reactions that sustain a cell.

Mitochondria Organelles inside eukaryotic cells that mainly perform respiration.

Morphology Shape of the body or body part.

Mutations Changes to the sequence of letters in nucleic acids.

Natural selection The process that drives adaptation through survival of the fittest.

Nucleic acids Molecules of DNA or RNA.

Nucleotides The building blocks of nucleic acids, each with one of four chemical 'letters' or bases.

Nucleus The organelle in eukaryotic cells that contains DNA and intitiates gene expression.

Organelles Little 'organs' inside cells that perform at least one activity.

Organs Body parts in a system that performs an activity, such as digestion or reproduction.

Phenotype Characteristics that result from a genotype.

Physiology The processes that sustain a body.

Polypeptides Molecules encoded by genes, each made from a string of amino acids.

Prokaryotes Organisms usually consisting of a single cell with DNA naked in the cytoplasm.

Proteins Folded structures that perform functions in cells, made from one or more polypeptides.

Recombination Creating new gene combinations by swapping nucleic acid between sections of chromosomes.

RNA Ribose nucleic acid, a molecule with four bases (U instead of DNA's T) that often occurs as single strands.

Selfishness Behaviour of an individual or gene that appears to act in its own interests.

Species A distinct type of organism, often interpreted as a population whose members can interbreed.

Symbiosis A close relationship between two species that often benefits at least one partner.

Tissues Groups of cells that perform a task, such as movement or communication.

Index

About the author

JV Chamary is an award-winning science writer. He studied biology at Imperial College London and has a PhD in molecular evolution and genetics from the University of Bath.

Greenfinch,
An imprint of Quercus Editions Ltd
Carmelite House
50 Victoria Embankment
London EC4Y 0DZ

An Hachette UK company

First published in 2015

A CIP catalogue record for this book is available from the British Library

PB ISBN 9781529438437
eBook ISBN 9781529440133

Quercus Editions Ltd hereby exclude all liability to the extent permitted by law for any errors or omissions in this book and for any loss, damage or expense (whether direct or indirect) suffered by a third party relying on any information contained in this book.

SRD

Printed and bound in India by Manipal Technologies Limited, Manipal

MIX
Paper | Supporting
responsible forestry
FSC™ C104740
www.fsc.org

Papers used by Greenfinch are from well-managed forests and other responsible sources.